SEAHORSES

SEAHORSES

A Life-Size Guide to Every Species

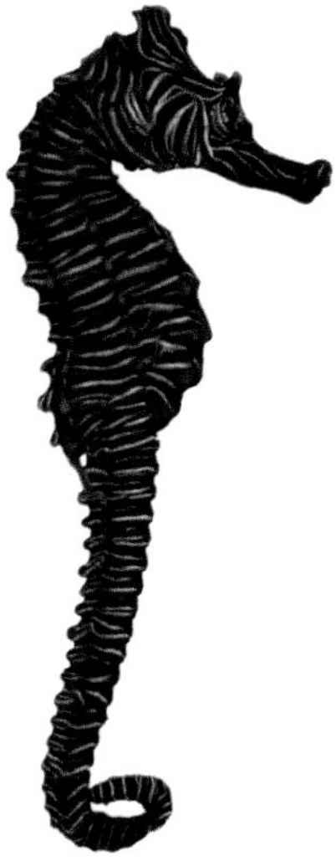

SARA LOURIE

THE UNIVERSITY OF CHICAGO PRESS

Chicago and London

In memory of Denise Tackett
her love of seahorses and of the underwater world

Sara A. Lourie is a research associate with Project Seahorse. She has identified multiple new seahorse species and is the author of several books and articles on seahorse taxonomy.

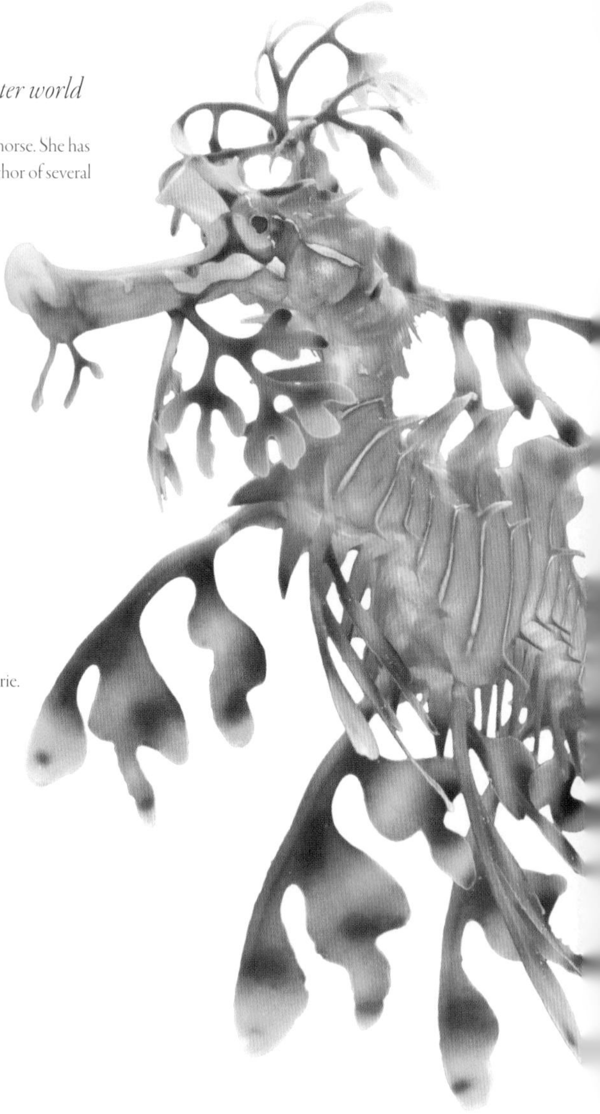

The University of Chicago Press, Chicago 60637

Printed in Malaysia

33 32 31 30 29 28 27 26 25 24 2 3 4 5

ISBN-13: 978-0-226-33841-5(cloth)
ISBN-13: 978-0-226-33855-2(e-book)
DOI: 10.7208/chicago/9780226338552.001.0001

Library of Congress Cataloging-in-Publication Data
Lourie, Sara A., author.
Seahorses : a life-size guide to every species / Sara Lourie.
pages cm
Includes bibliographical references and index.
ISBN 978-0-226-33841-5 (cloth : alkaline paper)
ISBN 978-0-226-33855-2 (e-book) 1.
Sea horses. I. Title.
QL638.S9L68 2016
597'.6798—dc23 2015035760

This book was conceived, designed, and produced by
The Bright Press, an imprint of the Quarto Group
1 Triptych Place, London
SE1 9SH, United Kingdom
www.quarto.com

PUBLISHER Susan Kelly
CREATIVE DIRECTOR Michael Whitehead
EDITORIAL DIRECTOR Tom Kitch
ART DIRECTOR Wayne Blades
EDITOR Jamie Pumfrey
COMMISSIONING EDITOR Kate Shanahan
DESIGN JC Lanaway
PICTURE RESEARCH Alison Stevens

MIX
Paper | Supporting responsible forestry
FSC® C007207
FSC www.fsc.org

CONTENTS

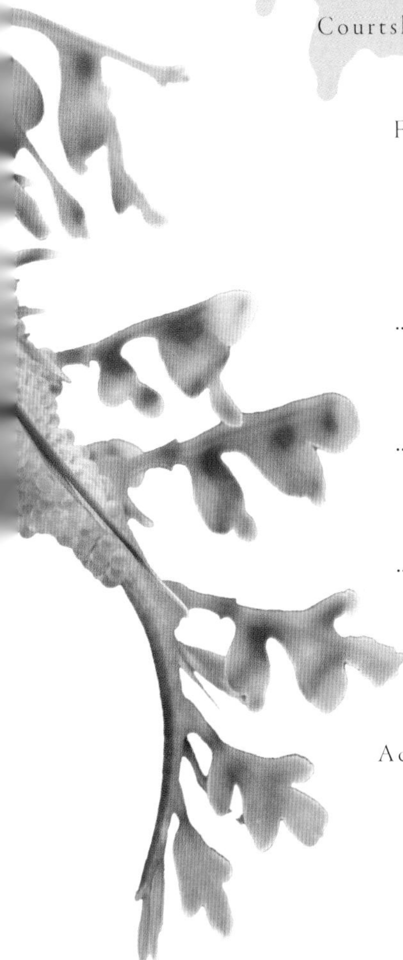

INTRODUCING SEAHORSES

WHAT ARE SEAHORSES?

HORSES OF THE SEA

Seahorses are among the most charismatic marine creatures. They are so unlike typical fishes that they were once considered to be marine insects. With their upright posture, strange body-shape and features, and incredible camouflage, it is no wonder that people used to think that seahorses were mythological beings and assigned them such roles as bearing sea nymphs and pulling Poseidon's chariot. The scientific name for the seahorse genus, *Hippocampus* (in this book shortened to *H.*), is derived from the Greek words *hippo* (meaning horse) and *kampos* (meaning sea monster).

Today we know that seahorses are in fact real fish, albeit most unusual ones, and that there are at least 40 different species. Like other fish, they have gills, fins, and a swim bladder. However, they swim vertically, propelled by a fin on their back, unlike typical fish, which swim horizontally and move by waving their tail fin. Seahorses lack scales—instead their bodies are covered by skin stretched across bony plates. The junctions of these plates protrude into ridges, bumps, and spines, the form of which can help distinguish the different species. They have prehensile tails that can literally hold your hand (or at least your finger), and males possess a marsupial-like pouch within which the young develop. Their curious lobe-like gills resemble miniature bunches of grapes, and they have eyes that can move independently of one another, like those of chameleons. The seahorse snout is tubular, with a tiny toothless mouth at its tip, through which the animal sucks hapless prey (mostly small crustaceans) from the water column or off the surface of vegetation. Food passes through a short and simple gut that lacks even a differentiated stomach.

Perhaps one of the most remarkable things about seahorses is the fact that it is the male who bears the young. Female seahorses still produce the eggs, and the male produces the sperm. However, when the eggs are ripe, the female deposits them in the male's brood pouch, and she plays no further role in the development of the young. The male, meanwhile, fertilizes the eggs inside his pouch, and undergoes an extensive pregnancy during which the young are nourished by the yolk-sac provided with each egg, bathed in a liquid that gradually changes from being like body fluids to resembling seawater over the course of a pregnancy. The seahorses grow until they are ready to be released into surrounding waters.

Seahorses are found in all tropical, subtropical, and temperate waters, meaning that extensive dispersal must have occurred over time, despite lacking a long planktonic dispersal phase. How that dispersal has been achieved may have something to do with the young grasping floating mats of vegetation with their prehensile tails, and rafting to far-off places. These secrets are starting to be unraveled as scientists begin to decode seahorse genetics.

As well as their strange shape and their unusual biology, seahorses are also highly prized for medicinal purposes, particularly in Asia.

An estimated 15–20 million seahorses are traded every year to satisfy the demand for traditional Chinese medicine. Live seahorses are also traded for aquaria, and dead (dried) ones sold as souvenirs. This exploitation rate, coupled with extensive trawling and destruction of their habitats has raised conservation alarm bells worldwide. Seahorses were among the first marine species to be listed by the Convention on the International Trade in Endangered Species (CITES). Many countries have national laws that help protect them, and volunteer divers in a number of countries have been recruited to help collect information on seahorse populations in the hope that this information and heightened awareness of the plight of these creatures will help save them.

It is becoming increasingly clear that the world beneath the waves (more than 70 percent of the Earth's surface) is suffering hugely from

SYNGNATHID EVOLUTION

Hypothesis of evolutionary relationships among a selection of Syngnathids. Based on work by Hamilton, Wilson, Teske and others. Note that the precise arrangement of taxa is uncertain in some cases, and that precise dates of divergence between taxa are not well known. Dark green lines show relationships or times that are better supported than light green lines. Dotted lines show approximate dates of divergence or fossil ages.

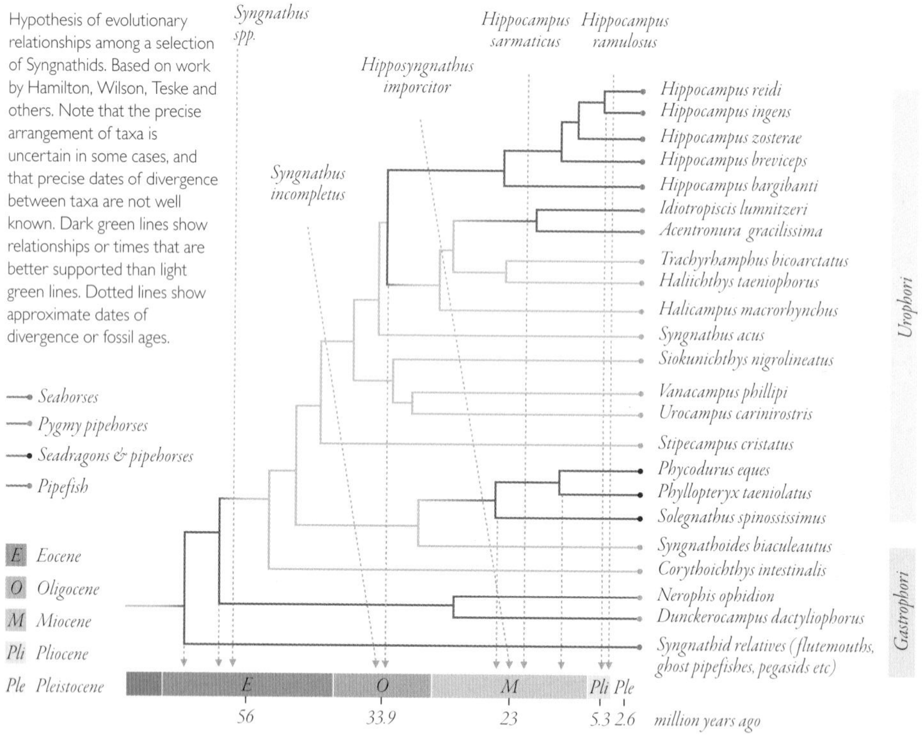

overexploitation, habitat destruction, pollution, and climate change. We can no longer claim complete ignorance of what is going on, and continuing with the status quo will only lead down a path of further destruction and disruption of the biosphere. However, if we can mobilize efforts to save the charismatic seahorse—and who doesn't love seahorses?—it would go a long way toward addressing many of the issues that are causing problems in coastal areas of the world's oceans.

THE FAMILY SYNGNATHIDAE

Seahorses are members of the teleost (bony fish) family Syngnathidae, most commonly known as pipefish. The family name, derived from Greek, refers to these fishes' tube-like snouts (*syn* = joined/fused, *gnathos* = jaws), and the family shares other similarities such as a bony external skeleton, reduced number of fins, and male brooding of the eggs (and young in seahorses). Within the Syngnathidae there are a variety of body forms, ranging from straight, unornamented pipefish, to half-bent pygmy pipehorses, to fully curled seahorses and highly ornamented and camouflaged seadragons. Surprisingly, however, these features do not necessarily mirror their evolutionary relatedness. For example, the seahorses (subfamily Hippocampinae) are closely related to pygmy pipehorses, and also to pipefish (such as *Trachyrhamphus*) that are straight and look very "normal." Genetic research has shown that some features, such as a grasping (prehensile) tail, or dermal flaps that increase a species' camouflage, may have evolved independently in different groups on a number of separate occasions (known as convergent evolution) and do not necessarily reflect a single common ancestor.

Within the family, the major division is between those species that brood their eggs on the front of their trunk region (the Gastrophori) and those that brood them on the underside of their tail (the Urophori). This distinction was first described by Herald in 1953, but has recently been supported by genetic evidence. Seahorses, pipehorses, pygmy pipehorses, and seadragons along with many straight pipefishes all fall into the second group, while the flag-tail pipefish and others that are straight and lack tail fins are in the first. Overall there are about 230 species of pipefishes distributed across 55 genera, however, this figure is only an estimate since there has been little taxonomic research done on the group since Dawson published his book on Indo-Pacific pipefishes in 1985.

The family Syngnathidae is a part of a higher order group, or order, called Syngnathiformes, which includes other families such as the Fistulariidae (cornet fishes), Aulostomidae (trumpet fish), Solenostomidae (ghost pipefishes), and Centriscidae (bellows fish). The majority of the species in these families occur in the Indo-Pacific, particularly in the Coral Triangle area, between the Philippines, Indonesia, and Papua New Guinea, as is the case for many other marine families.

MYTHS AND LEGENDS

The earliest depictions of what could be considered a seahorse (or a close relative such as a Ribboned Pipefish) were on cave walls in Arnhemland, northeast Australia. These images may have represented the Rainbow Serpent, or creator god, of the Aboriginal Dreamtime. The timing of their execution (about 6,000–8,000 years ago) coincided with rising sea-levels at the end of the last ice age, and it has been suggested that it was the encroachment of the sea, and flooding of the shallow continental shelf, that gave people increased opportunities to witness marine life of the shallows (such as seahorses), which in turn gave artists their inspiration for the Rainbow Serpent.

On the opposite side of the world, seahorse-like creatures were frequently depicted in Roman, Greek, and Etruscan mythology and art from the 6th century BCE onward. The Hippocampus (or Greek ιππόκαμπος) was a mythological creature that was half horse and half fish, possibly inspired by real seahorses. A pair of Hippocampi drew the chariot in which Poseidon (the Greek god of the sea) or Neptune (his Roman counterpart) rode. Hippocampi were often steeds for nereids (sea goddesses) or Triton (the son of Poseidon and Amphitrite), and they frequently adorned tombs, amphorae, and coins. A hoard of 30 silver coins was discovered in the Yizreel Valley in Lower Galilee in 1981. These are believed to have been minted in the 4th century BCE, and showed the local god, Melqart, riding on a Hippocampus. In Greece, a Hippocampus statue, topping the Temple of Poseidon in Helikos, apparently caused damage to fishermen's nets for many years following the submergence of the city in about 373 BCE (as a result of a tidal wave and ground liquefaction). Mosaics portraying Hippocampi decorated the Roman Baths at Aquae Sulis in Bath, UK, and in Rome, and Hippocampi are represented on the Great Dish (or Neptune's Dish) that forms part of the 4th century Roman Mildenhall Treasure, found in Suffolk, UK, in 1942.

The meaning of funerary art incorporating Hippocampi has been the subject of much debate. It has been suggested that Hippocampi had a role to play in conveying the dead on their watery journey to the Underworld. However, some Etruscan, Roman, and Lucanian depictions suggest an alternative interpretation. The armed warriors in these portrayals, apparently doing battle with Hippocampi, suggest that Hippocampi were in fact monstrous obstacles that the soul must overcome on its way to the Afterworld.

Discovered near Sardis (in today's Turkey), a tiny gold Hippocampus brooch became the center of a great controversy over illegal looting of burial mounds. It was part of an incredible Lydian hoard, from the 6th century BCE, that ended up in the Metropolitan Museum of Art in New York, after being smuggled out of the country and sold by unscrupulous dealers in the early 1960s. The full story did not come to light, however, until the 1980s, after which

ABOVE Hippocampi were depicted in mosaics and sculptures as creatures that were half horse and half sea serpent. *The Triumph of Neptune* is a marble and glass mosaic from Tunisia in the 3rd century AD.

Turkey sued the museum for its return. The case seemed to come to a conclusion in 1993 when the Hippocampus and other artifacts were returned to Turkey. However, the Hippocampus was stolen again in 2006, since when its whereabouts have been a mystery. The brooch now on display at the Usak Museum (near where the hoard was found) is a replica, and the original may no longer even exist.

Other times and cultures have also incorporated the strange-looking seahorse into their mythology. A male god holding two seahorses is found on one of the outer plates of the 1st century BCE Gundestrup Cauldron, the largest surviving European Iron Age silverwork, found in a peat bog in Himmerland, Denmark, in 1891. Stone carvings from the 3rd and 4th century in Scotland may represent seahorses, or loch-dwelling water kelpies. In 2009, an Anglo-Saxon seahorse was discovered among the intricately filigreed weapon decorations found as part of the 4,000+ item Staffordshire Hoard. Only about 1½ inches (4 cm) tall, this piece shows a complexity and intricate craftsmanship that defies belief—three of the tiny metal spirals fit into the length of a single grain of rice.

During Medieval times, Hippocampi appeared on heraldic charges. Hippocampi persisted through the Renaissance and Baroque Periods, adorning paintings, and statues. Britannia (the female personification of Britain) is pulled through the waves by a team of strong (sea) horses on British stamps issued between 1913 and 1939, as well as on medals for naval service to Britain. Finally, a winged Hippocamp (*hippocampe ailé*, affectionately referred to by employees as *la crevette*) was adopted in 1933 as the logo for Air Orient (and its successor Air France). The front half of the *hippocampe ailé* represented Pegasus, the Greek winged horse (in reference to speed and air), and the back half represented the Dragon of Annam (depicting the link with French Indochina and, being fish-like, the company's use of flying boats). Another story is that the original idea for the *hippocampe ailé* came from the founder of Air Orient's crash into the Bay of Naples (full of seahorses) in the 1920s, though this may, or may not, have any truth to it.

In Chinese cultures, seahorses were believed to be some kind of sea dragon. They were revered as symbols of power and good luck. These attributes have continued to be applied to them, and no doubt contribute to the medicinal value they are believed to possess. The good luck they are said to bring makes seahorses popular among seafarers. Captain Cook's second ship, *The Resolution*, had a seahorse as a figurehead, and fishers in Southeast Asia commonly keep dried seahorses as protection and to bring them good luck in their fishing expeditions.

The symbolic meanings and values associated with seahorses reflect their biology and nature, and are said to include high perception, patience, protection, inflexibility, persistence, perspective, friendliness, contentment, and generosity, as well as imagination, creativity, good luck, fatherhood, vigilance, grace, and the power of the ocean.

Whatever symbolic, cultural, and mythological relationships humans have with seahorses, it is clear that they have captured people's imaginations for centuries. Today they appear in children's books (such as Eric Carle's *Mr Seahorse*, and Graham Base's *The Sign of the Seahorse*), and in role playing games (for example the Pokémon seahorse character, Seadra). Two commemorative coins displaying seahorses (from Canada and the British Virgin Islands) were minted in 2014 and seahorses have been portrayed on many postage stamps (from nations including Palau, USA, Vietnam, Bermuda, the Philippines, and Bulgaria). Seahorses are also charismatic icons and ambassadors for marine conservation. The hope is that people's love of seahorses is strong enough to motivate them to do what they can to help ensure that these remarkable animals (and their marine habitats) persist indefinitely into the future.

MORPHOLOGY

ADAPTATIONS OF THE HEAD

One of the most unusual features of seahorses is the fact that they orient themselves vertically in the water column, with their head bent at a right angle to their body so they are still looking forward. It is a non-trivial evolutionary change to have gone from a "normal" horizontal-swimming fish (like most pipefishes) to a vertical seahorse, and one that necessitates all kinds of anatomical modifications. In particular, changes to the spinal column are needed, and several of the vertebrae closest to the head have become modified so that the holes, through which the spinal cord pass, no longer line up, but instead are off-set in order to accommodate the right-angle bend.

The seahorse head has other unusual features. Most prominently, the snout is extremely elongated. From a morphological point of view it is not technically the jaw bones that are elongated, but other bones that are normally found toward the main part of the skull which have become extended during seahorse evolution. The jaws themselves are the tiny "lips" at the end of the snout, which open and close to allow the entry of small crustaceans and other prey.

The long, tube-like snout enables seahorses (and pipefish) to suck their prey rapidly out of the water. They are experts at the technique. *Hippocampus erectus* and *H. reidi* have been recorded as sucking up passing plankton in fewer than 6 milliseconds, and juveniles are even faster (2.5 milliseconds). This is the fastest recorded suction feeding among all teleost fishes. To provide this rapid suction, syngnathids have developed a specialized technique called

DIAGRAM OF A SEAHORSE SKULL

Franz-Odendaal, T & Adriaens, D. (2014) EvoDevo. 2014(5): 45

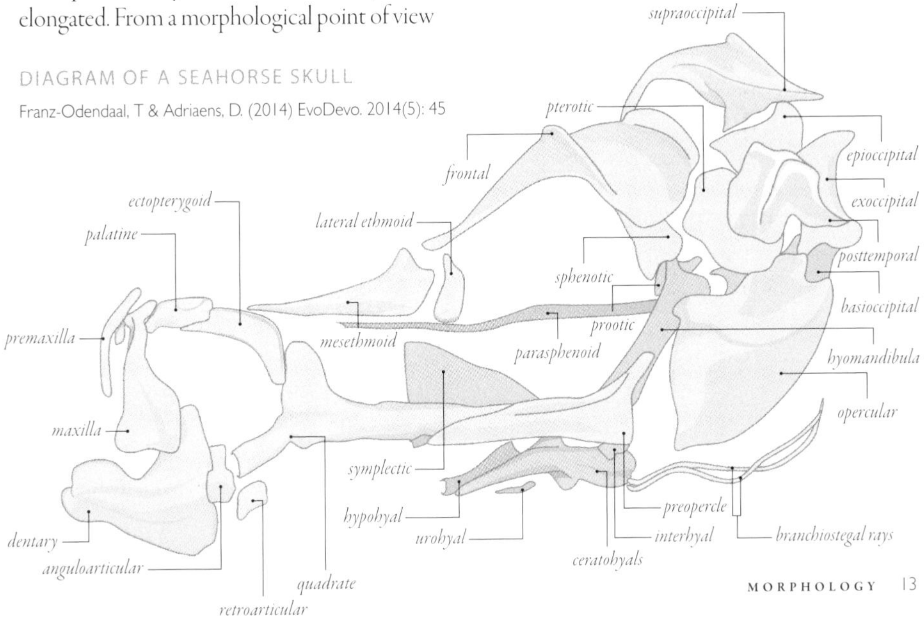

pivot-feeding. This is initiated by depression of the hyoid bone (or "trigger") under the snout, followed almost instantaneously by the rotation of the head, elevation of the front of the skull (neurocranium), and expansion of the cheeks. This causes a large pressure difference between the mouth cavity and mouth opening, and creates a flow of water into the mouth, along with the intended prey item. It also causes a large pressure differential between the mouth cavity and the gills, which in turn necessitates some special modifications of the gill cavity. One of these is the reduction of the slit-like opercular opening behind the gills (as found in most fishes), to a small hole (near the pectoral fin in most seahorses, but right behind the coronet in pygmy seahorses). Another modification is the extra strong bony gill cover, which is convex in shape, and thus able to withstand the pressure.

On top of the seahorse's head is a bony projection, called the coronet (or crown). This too may have a role to play in the feeding mechanism. Depending on the species, the coronet may be more or less developed (it is basically flat in *H. capensis* and *H. minotaur*). When the neurocranium is elevated during feeding it articulates with the bones that make up the front of the coronet and produces a clicking sound. Seahorses have been reported to produce clicking sounds at other times as well (for example, after being moved to a new tank, or when under stress), and it may be that clicking is not only a by-product of feeding, but also a means of communication.

Another interesting feature of the seahorse is its eyes, which move independently and alternately like those of a chameleon. This is of great advantage to an animal that is an ambush predator, relying on its camouflage to remain hidden and avoid detection, either by prey or by predators. It means that it can get a good view of the environment while keeping its head and body motionless. Seahorses have very acute vision, allowing them to locate and successfully strike at small, fast-moving crustaceans. They have a specialized part of the retina called the fovea that helps them do this. The funnel-shaped fovea has a much higher concentration of photoreceptor cells than does the surrounding area of the retina, and it is believed that it can magnify images, maintain accurate fixation, and act a directional focus indicator.

MAJOR FEATURES OF A SEAHORSE HEAD

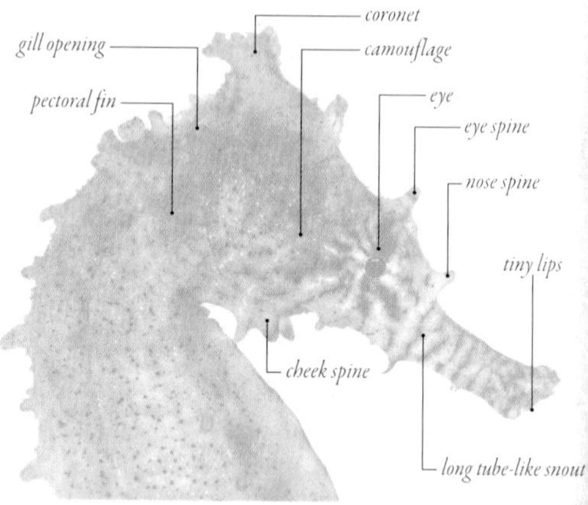

BONY PLATES

The bodies of seahorses (and relatives) are protected by an external armor of interlocking bony plates, as opposed to the scales of typical fish. These plates define their shape, limit their movement, and, most importantly, protect them from predation and make them more unappealing to potential predators in the first place. They interconnect via joints that slide past one another, and are frequently ornamented with ridges, indentations, furrows, or spines. In the trunk region seven plates interconnect to form a heptagonal shape (flat on the back with a central mid-line in the front). This switches to a square shape, made up of four interconnecting plates, on the tail. Each ring of plates is centered on a vertebra, and these are easily visible as a series of ridges that cross the body and tail. The number of ridges (or rings) on the trunk (TrR) and tail (TaR) are important taxonomic features that can be counted to help identify the species.

Seahorse plates are much reduced in comparison to those of pipefish. This allows seahorses a greater degree of mobility, particularly of the tail. The plates can be compressed up to nearly 50 percent before being permanently damaged. Even when compressed beyond 50 percent they do not become brittle, but tend to buckle providing seahorses with unparalleled protection. In fact, the biomechanics of seahorse armor is being actively researched to help inspire novel synthetic materials where stiffness, superior strength, and flexibility are all required.

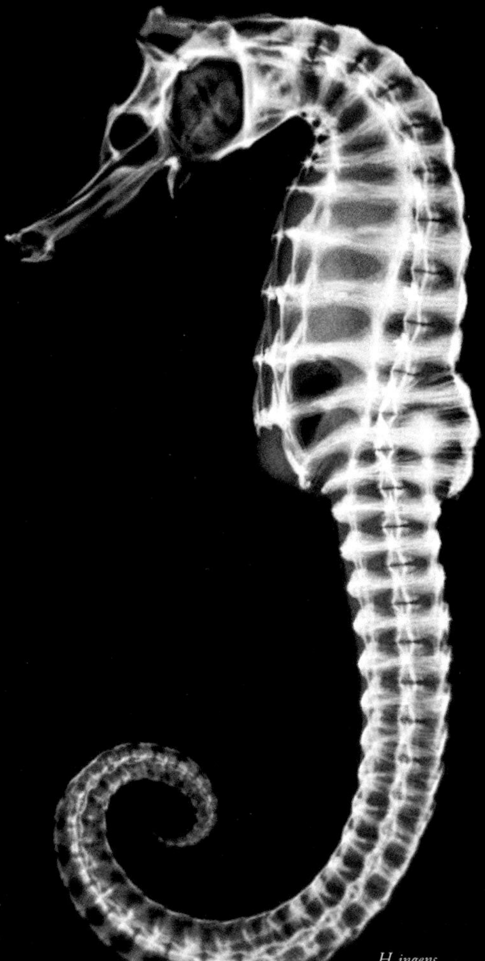

H. ingens

ABOVE A seahorse's bony plates are clearly visible in an X-ray, as are the vertebral column and the bones that make up the skull. Note that seahorses lack ribs.

ABOVE A seahorse's prehensile tail can be used to grip any suitable object, even by its very tip.

PREHENSILE TAIL

The grasping (prehensile) tail of the seahorse is a marvel of engineering. Encased by bony plates, yet incredibly flexible, it can bend forward (creating a spiral), backward (to a lesser extent), and sideways (somewhat). It enables the seahorse to grab onto a holdfast, such as a piece of seagrass, seaweed, coral, mangrove root, or something man-made, and thus anchor it against the moving water. A threatened seahorse, rather than swimming away, will instinctively hold tighter to its holdfast, and tuck its head to its chest. This is a problem when the holdfast is a fishing net, or a crab trap.

In order to curl around a holdfast, the series of bony plates that make up the tail are slightly deformable, and can also slide over one another. The prehensile action is mediated by two different kinds of muscle inside the tail. The hypaxial muscles which lie between

PREHENSILE TAIL

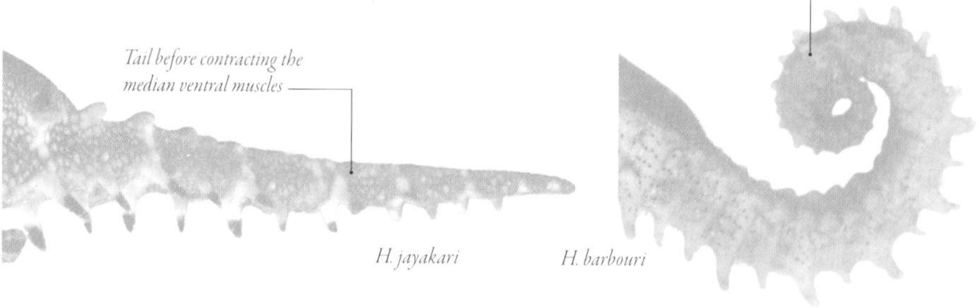

parallel sheets of connective tissue power fast grasping, whereas the median ventral muscles (which are believed to be modified fin muscles) are capable of sustained holding. This muscle structure is unique to seahorses.

Other taxa in the syngnathid family also have prehensile tails, including Alligator Pipefish (*Syngnathoides biaculeatus*), pipehorses (*Solegnathus* spp.), and pygmy pipehorses (for example, *Acentronura* spp.). Surprisingly, however, the trait seems to have evolved independently at least twice. Alligator Pipefish and pipehorses are more closely related to pipefish without prehensile tails than they are to seahorses and pygmy pipehorses. The different genera have come up with rather different solutions as to how to create a prehensile tail, exhibiting different muscle arrangements and different degrees of development of bony plates on their tail. For example, in *Syngnathoides biaculeatus* the first 70 percent of its tail is relatively rigid and only the last 30 percent lacks bony plates and is prehensile.

Solegnathus spp., that use their tails to grab onto sea whips, have plates along the entire length of their tail. For the last third of the tail these plates are no longer attached to one another, enabling greater movement. The pygmy pipehorses (for example, *Acentronura*) that are commonly found attached to vegetation are most similar to seahorses and show a much more seahorse-like arrangement of plates and muscles.

FINS

A reduced number of fins is one of the characteristics that unites seahorses and their relatives. Other fish have pelvic, pectoral, dorsal, caudal, and anal fins, but adult seahorses have only pectoral, dorsal, and anal fins (i.e. they have lost their paired pelvic fins and their caudal (tail) fin). Some pipefishes have caudal fins. All fins (except for the anal fin) can be folded flat, which may be advantageous in dense vegetation.

The dorsal (back) fin is the longest and is primarily responsible for providing forward motion. Like the other fins, it is supported by fin rays made of cartilage, and these are raised, lowered, and moved from side to side by a series of muscles called elevators, depressors, and inclinators. Because of their rigid body, seahorses move exclusively by undulation of their dorsal and pectoral fins. These can beat at a rate of more than 40 times per second (40 Hz), and up to 54 Hz in *H. zosterae*. Other fish that use their dorsal fins for propulsion typically beat them at around 2 Hz. This fin undulation produces a much less powerful form of motion than the typical side-to-side movement of most fish tails, and seahorses cannot swim very far or fast. However, this mode of propulsion is supremely well suited for high maneuverability in complex habitats (such as coral reefs and seagrass beds). The rapid fin undulation without any body movement also has the advantage of being very discreet, not attracting the attention of potential predators or prey, which is very useful for animals that rely on their camouflage as protection and for effective ambushing of prey. In fact, the frequency of dorsal fin undulation in seahorses is so high that it exceeds the point of "flicker fusion" in humans (and likely in predators too), meaning that its movement is too fast even to be detected by an observer's eye.

The two pectoral fins, located on either side of the neck just behind the gill cover (operculum), act as stabilizers and help the seahorse steer as it moves around its habitat. They also provide some forward thrust. The anal fin is tiny, usually with only two fin rays supporting it. This is found just above the anal opening, and may serve a hygienic function, have a role in mating or expelling the young at birth, or may simply be a vestigial remnant of a more developed fin.

Some species (for example *H. kuda*, *H. mohnikei*, and *H. spinosissimus*) have been found to have a tiny caudal (tail) fin when they are first born, reflecting their evolutionary relationship to the tailed pipefishes. However, this is lost within a few days and adults are left with a naked, finless tail adapted entirely for grasping.

MORPHOLOGICAL FEATURES

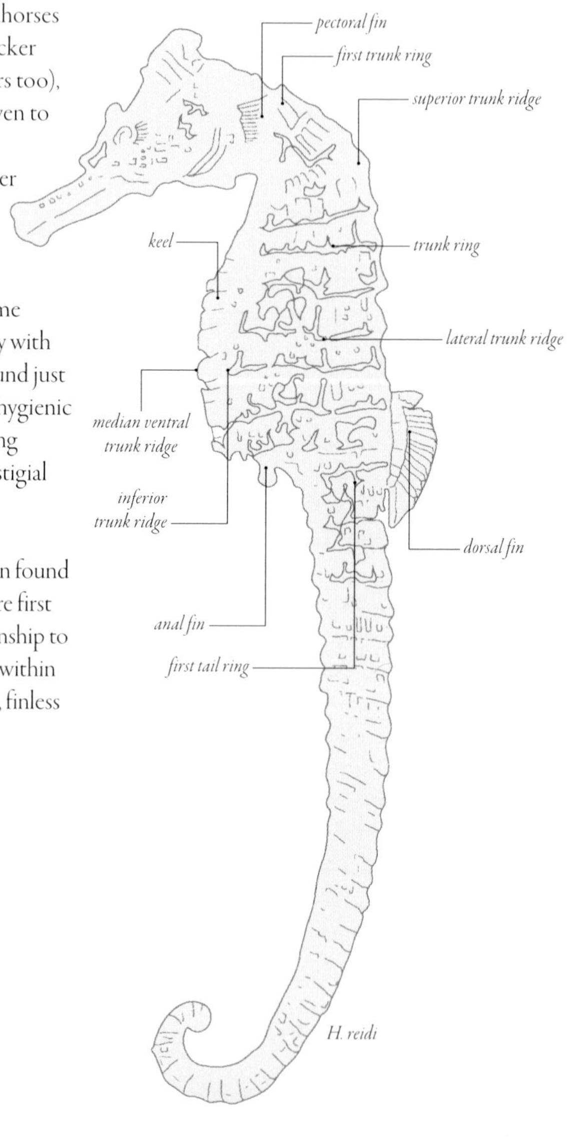

RIGHT External morphological features of a typical seahorse body (see p. 14 for features of the head).

POUCH

Male seahorses are the ones that look after the eggs and developing embryos, and they do so in a special sac attached to their tail called a pouch. Seahorse relatives show various degrees of development toward the fully enclosed pouch seen in seahorses. Some species, such as the seadragons (for example, *Phycodurus*), the Straightnose Pipefish (*Nerophis ophidion*) or flagtail pipefishes (for example, *Doryrhamphus*) simply glue the eggs onto the underbelly of the male. Others, such as *Corythoichthys*, have flap-like extensions, or pouch plates, on either side of the eggs to protect them. Skin folds that are large enough to overlap one another occur in *Syngnathus*. Similar folds, that cross over to such an extent that they actually separate the brood into left and right halves, occur in *Pseudophallus*, and flaps that completely join at the posterior two tail rings are found in *Amphelikturus*. It is just a short evolutionary leap from there to a fully sealed pouch, as is typical for seahorses, with only a small opening near the top end to allow the deposition of the eggs, and the release of the young. Among seahorses, only the pygmies (*H. bargibanti*, *H. colemani*, *H. denise*, *H. pontohi*, and *H. satomiae*) lack the pouch altogether, and the young appear to be brooded within the body cavity itself. Interestingly, despite its small size (similar to other pygmies), *H. debelius* appears to resemble more closely the larger species in this regard and does have a defined brood pouch on its tail.

Whether the brooding area is found on the body (trunk) or on the tail has been shown to be an important distinguishing feature among syngnathids. Those species that brood on the trunk are called Gastrophori (literally "stomach carriers"), whereas those whose brooding structure is on the tail are called Urophori (or "tail carriers"). Seahorses have evolved from the Urophori line. Genetic data support the hypothesis that syngnathids can be split into two separate main lineages, defined by the position of their brooding area. This key division apparently happened early on in syngnathid evolution, and both lineages have subsequently undergone major radiations into multiple species.

The seahorse pouch is a heavily vascularized container into which the female deposits her ripe eggs. The lining of the pouch changes significantly during the breeding cycle. Once in the pouch, the eggs become embedded in small cups in the pouch wall, where they are bathed in a liquid that initially has a chemical make-up resembling internal body fluids. Over the course of the pregnancy, its composition gradually changes to become similar to seawater. It is unclear to what extent the male provides nutrition to the developing young in seahorses. The inner surface of the pouch contains secretory cells (modified flame cone cells) that modulate the composition of the placental fluid, and may also stimulate the breakdown of the outer layer (chorion) of the eggs to provide nutrition to the young. The eggs come supplied with their own yolk sacs, which provide the developing embryos with nutrition. The father plays an active role, however, in providing minerals, aeration, and removal of wastes. That this can be considered a real pregnancy is attested to by the fact that the pouch is a physiologically complex organ in which the young have all their needs met. The complex activities are mediated by the same hormone, prolactin, as mediates human pregnancies.

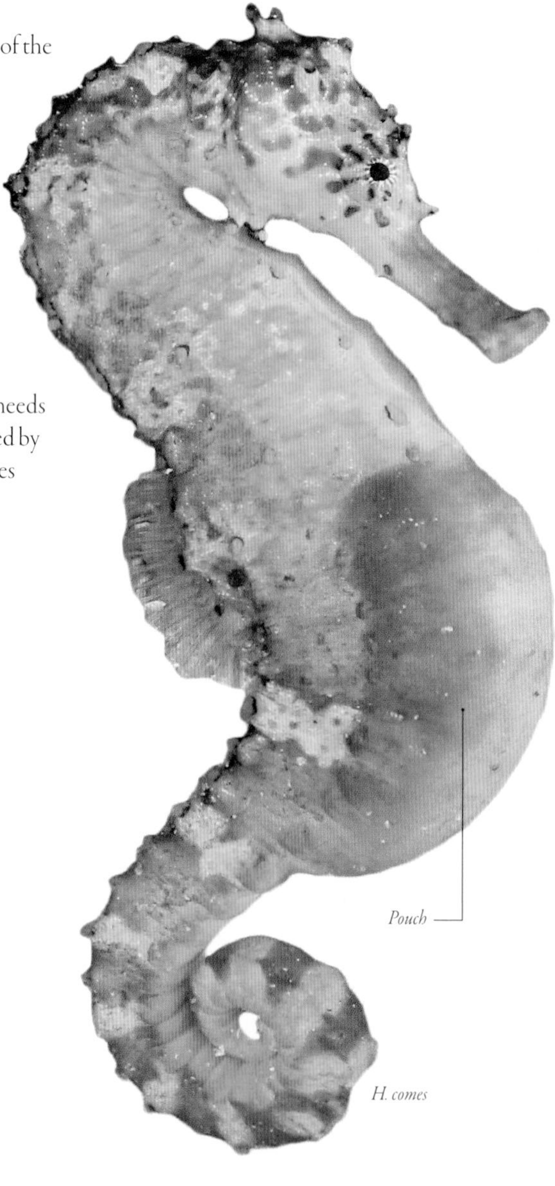

RIGHT A heavily pregnant male *H. comes* displays his distended belly as he swims through the water. Male seahorses typically have smaller home ranges than females, perhaps due to the encumbrance of their huge pouch when pregnant.

INTERNAL ORGANS

Internally, seahorses have basically the same organs as other fishes. They regulate their buoyancy by means of a gas bladder, and they have a heart, liver, kidneys, and gills.

The seahorse gas bladder does not have a direct connection with the external environment (i.e. it is physoclistous). It is relatively large and centrally located, to offset the weight of the animal's bony external skeleton. The volume of gas within the bladder is regulated by a heavily vascularized oval-shaped area (the gas gland), which can secrete or reabsorb gas as needed, allowing the seahorse to maintain neutral buoyancy. Diseases, or (in captive animals) imbalances in the aeration of aquaria where seahorses are housed, can cause problems with over- or under-inflation of the gas bladder, and hence buoyancy difficulties. Additional gas-related problems include the catch-all term "gas-bubble disease" in which bubbles of gas appear either under the skin, or within the body cavity, and pouch emphysema where gas is trapped within the pouch. All of these conditions are potentially lethal.

The seahorse heart is small and located in the head region next to the gills. Blood from the heart is pumped to the gills, where it is aerated before continuing on to the rest of the body. The gills are often (erroneously) referred to as "tufted." They are in fact similar to those of other fish, with five gill arches (which house blood vessels), to which are attached rows of gill filaments. At right angles to the filaments are flaps (lamellae) covered in tiny projections (microvilli), which increase the surface area of the parts where gas exchange takes place. The reason that syngnathid gills look different from those in other fish has to do with the extremely small space in which they must fit. As a result, the gill filaments are short and broad (compared to being longer and more filament-like in other fish), and when viewed from their ends look almost pear-shaped. Compared to other fish, seahorses have relatively few gill filaments, however, this is sufficient to provide adequate oxygen for their relatively sedentary life-style. The gill cover (operculum) is convex to withstand high internal pressure (caused in part by their suction feeding), and the movement of water over the gills is achieved by an opercular pump mechanism.

The feeding mechanism of seahorses has been explained (see Adaptations of the head, p. 13), however, mention must be made of the rest of the seahorse digestive system. Seahorses lack teeth (19th-century newspaper advertisements for "seahorse teeth" refer not to *Hippocampus,* but to *Hippopotamus,* whose teeth were prized for their ivory), and they have no true stomach. Food is sucked whole into the mouth cavity, and then passes down through the esophagus into a simple tubular foregut. This is separated from the mid-gut (or intestine) by a slight constriction that equates to the pyloric sphincter in species that have a stomach.

The entire length of the digestive tract in seahorses is only about 40 percent as long as the whole fish (i.e. relative gut length ≈ 0.4). Typically carnivores have relative gut-lengths of < 1.0, whereas herbivores and detritivores have relative gut-lengths > 3.0. Seahorses are carnivorous and they are not very active. Both of these factors may account for why they have evolved not only a very short gut, but also no stomach. The consequence of this limited digestive system is that seahorses need to spend most of their time feeding, and they frequently have difficulty digesting non-natural food (when in captivity) because they lack the digestive enzymes that would normally be produced by the stomach. The liver does produce bile that helps with digestion, and it also helps with detoxification of the blood.

Undigested waste materials pass through the gut and out of the anus, located immediately above the pouch opening in males, and below the ovipositor in females. Other wastes are filtered from the blood into urine by the kidneys. These are elongated paired organs that lie against the spinal column. All seahorses and pipefish studied to date (> 29 species) have tubule-dominated kidneys. They lack the "tufts" of capillaries (glomeruli) that are usually nestled within cup-like sacs (Bowman's capsules) and form the basic units of most vertebrate kidneys. The "aglomerular" condition in syngnathids is considered to be a water- and energy-saving adaptation for animals living in saltwater, particularly those that are very sedentary.

CAMOUFLAGE

Unlike other species that may use speed to escape predation, seahorses depend on cryptic behavior and camouflage as their primary defense mechanisms to avoid becoming someone else's dinner, as well as a method to surprise their own prey. Their irregular body shape, spines, and bumps help break up their outline, so that they blend into the background, and many have patterns, spots, and mottling that further disguise them. Their masterful disguises also make finding them for a field study quite a challenge.

Some seahorses have spots that may provide a warning function. For example, the large dots on the neck and dorsal surface of *Hippocampus trimaculatus* (and *H. camelopardalis,* and *H. planifrons*) look like the eyes of a crab when the seahorse is bent over.

In addition to color and markings, seahorses can further camouflage themselves by allowing algae, bryozoans, and small hydra to attach to their body surface. It appears that special cells on the body surface (epithelial flame-cone cells)

OVERLEAF Pygmy seahorses, such as this *H. denise*, are particularly good at matching the color and texture of their seafan habitats. This individual is a male ready to give birth.

encourage the attachment of these growths, by secreting a mucus cap to which these small animals and plants become attached.

Perhaps the most perfect example of camouflage is exhibited by the pygmy seahorse *H. bargibanti*. Bony protuberances (tubercles) on its body match the closed polyps of the gorgonian seafan on which it lives so precisely that the first specimens were not noticed until the seafan that they called home was collected and brought into an aquarium.

Most species of seahorses come in a range of colors, from black to brown, white, yellow, orange, and red. Since they are so variable in this regard, color is not a good way to identify species. Individual seahorses can even change color in order to blend in with their surroundings better, switching from black to neon orange in only a few hours or a few days. The mechanism by which they assess the color of their environment and match it with their own colors is unknown.

LIFE HISTORY AND BEHAVIOR

LIFE-SPAN

Seahorses live from less than one to 12 years, with most species living for three to four years. The young are already fairly advanced by the time they leave the pouch, and develop rapidly after birth, usually maturing in less than a year. Large species tend to live longer than smaller ones. In general seahorses are considered to be "r-selected," meaning that they mature quickly and produce relatively large numbers of offspring, as opposed to "k-selected" organisms, which mature slowly and produce few young (like many sharks).

Mark-recapture studies on *H. whitei* in New South Wales, Australia, have shown that this species lives for at least five years in the wild, although in captivity it rarely lives more than three; *H. guttulatus* also lives for at least five years in the wild and about 12 in captivity. The life-span of other species is shorter. Most individuals of *H. zosterae* will not reach their second year. Nevertheless, they mature quickly and are ready to breed at two or three months of age (depending on the temperature). It is possible for this species to have three complete generations in the space of a single year.

HOLDFASTS

One of the unique characteristics of seahorses is their monkey-like prehensile tail, with which they typically attach themselves firmly to a holdfast (usually an object rooted to the ground). Depending on their habitat, seahorses will hold onto a variety of things. These include seagrass blades, corals, sponges, seafans, mangrove roots, and even man-made objects such as discarded fishing gear, trash, protective swimming nets and jetties. Often they will simply choose whatever is available. However, some seahorses can be very particular about their holdfasts. One *H. comes* individual was found on the same piece of coral every time it was observed over a period of two years, and individual *H. guttulatus* specimens often stay attached to the same piece of seagrass for months. Other seahorses (for example *H. whitei* in Port Stephens) show a definite preference for sponges and soft corals (*H. dendronephthya australis*) even though their habitat is dominated by seagrass. Some species are extremely specific in their choice of holdfast, particularly the pygmy seahorses. *H. bargibanti*, for example, is found only on gorgonian seafans of the genus *Muricella*. They have perfected their camouflage to such a degree that they are almost impossible to spot. Their background body color is grayish-blue and slightly striated, just like the branches of the coral, and their body tubercles are large, round, reddish-pink or orange, and look just like the closed polyps of the coral. The orange *H. denise* is also a habitat specialist, matching its gorgonian coral hosts such as *Annella reticulata* and *Villigorgia* species.

Not all holdfasts are static. *H. spinosissimus* frequently use pencil urchins as holdfasts. Since the urchins move around, the seahorse gets a free ride. Imprudent seahorses that choose to attach to fish traps or crab or lobster pots undoubtedly get a nasty surprise when the owner of the traps comes to collect their catch. As a fisher in Vietnam was heard to comment during a discussion about seahorse conservation: "How can we protect a fish that is so stupid that it attaches itself to fish traps?"

Holdfast specificity can be a disadvantage for seahorses, particularly in habitats that are subject to disturbance. Either the seahorse is carried off along with its holdfast (since they tend to hold tighter to their holdfast when threatened), or it takes a while to find a new holdfast. Swimming while looking for a new holdfast can be a dangerous time. Several seahorses that were temporarily moved as part of a research project on *H. whitei* were observed to be eaten or attacked by predatory fishes while returning to their holdfasts, testimony to their poor swimming ability. Seahorses that are displaced have an impressive homing capability. One *H. guttulatus* apparently returned to its

OVERLEAF A juvenile seahorse remains pelagic (swimming in open water) for several days to a few weeks, during which time they can disperse.

original holdfast 492 feet (150 m) away after eight days, and another moved 197 feet (60 m) in one day to return home.

In areas where the water is relatively still, some seahorses will not bother with a holdfast, and instead simply curl up in a depression in the sand, or under an upturned piece of rock or trash. In the open water some seahorses attach to floating algae and in this way may be transported hundreds of miles. In some cases, this type of dispersal can lead to the founding of an entire new population of seahorses, especially if the rafting seahorse happens to be a pregnant male. Seahorses can feed on crustaceans associated with these floating mats of algae, and it is possible that a seahorse could complete its entire life cycle in floating algal habitats such as are found in the Sargasso Sea.

BELOW Seahorses are sit-and-wait predators that attach to holdfasts with their prehensile tails to anchor themselves against the water current.

HOME RANGES

Although not territorial in the sense of actively defending an area against intruders, seahorses do usually stay in a specific patch of habitat—their home range. This favored area may range in size from a part of a single seafan (for example *H. bargibanti*) to an area tens of square feet or meters in size (for example *H. whitei*). Even within this home range, individuals often focus their activity around a specific place. For example, individuals of *H. denise*, while roaming over much of a seafan during the day, will retreat to the same spot on the fan every night to sleep. This area is called the "core area" and is also the place where reproductive activities (greetings and breeding) take place. Both members of a breeding pair will tend to use the same core area.

Home ranges of different individuals may overlap. This seems to be particularly the case in *H. breviceps*, *H. capensis* and *H. guttulatus*. Having a specific home range, and a particular partner, presumably maximizes a seahorse's reproductive potential. Once they have established their pair bond and home turf they do not have to waste time looking for a partner in order to breed. Adult male *H. guttulatus* apparently do not tolerate other breeding males in the same territory, but they do tolerate juveniles and subadults. Females do not seem concerned if other females share the same living space, regardless of their age or reproductive status.

Within a species the size of home ranges can vary, with males tending to have smaller ranges than females, perhaps as a result of increased energy expenditure to move around when pregnant. In *H. whitei* the males typically spend their whole day within an area of only about 11 square feet (1 m^2) whereas females will range much more widely (110 square feet, 10 m^2).

Other species have larger territories, or possibly no territory at all. *H. abdominalis*, for example, typically roams over several thousand square feet or metres, and *H. spinosissimus* has been shown to favor motile pencil urchins as holdfasts, so they move as their holdfast moves across the sea floor.

FEEDING

Seahorses may not seem like typical scary predators, but they are in fact voracious eaters. In a single day, a seahorse may eat several thousand small prey items. A gut content analysis of juvenile *H. zosterae* found up to 200 items in the gut at a time. Seahorses particularly like tiny crustaceans, such as amphipods, copepods, mysids (shrimp), and crab larvae. However, they will also eat other crustaceans, small fish, eggs, larvae, snails, and other live prey. They have even been reported to be cannibalistic, with dropped eggs and even seahorse young in their guts. Basically, anything small enough to go into their snout is fair game (and sometimes they will even try something that is far too large to fit). Even pieces of vegetation have been found inside seahorse guts, although it is not known whether they have deliberately eaten these, or whether they were just accidentally ingested as the seahorse sucked up its prey.

Seahorses are visual ambush predators. Their prey is either floating (planktonic), on the surface of vegetation (epibiotic), or on the sea floor (benthic). They rely on their superb camouflage to avoid detection while they position their snouts beneath potential prey, and rapidly snap their heads forward and vacuum up their prey. The species *H. trimaculatus* has been observed to feed by forcefully squirting water into the sediment, and then sucking up the floating invertebrates.

The proportion of planktonic to epibiotic/benthic prey varies by species. For example, *H. guttulatus* typically sways back and forth with the water current while attached to a seagrass blade, or sits near the base of the seagrass, capturing planktonic prey as it passes, whereas *H. hippocampus* is a more active forager, typically snapping at vegetation to find its meals. Plankton feeders tend to have longer snouts than epibiotic/benthic feeders. A long snout increases the available volume of water that can be "hunted" and increases the efficiency of feeding. It can also increase the ability to forage in crevices. *H. zosterae*, which has a very short snout (as does *H. hippocampus*), was observed to snap more often at seagrass blades than the water column when feeding.

Seahorses lack teeth and have only a very simple digestive system, without a differentiated stomach or intestines. As a result, their digestion is not particularly effective and this may account for their huge appetite. Their extended parental care means that when seahorses are born they already possess a functional digestive tract and feeding apparatus, and they typically start feeding within the first hour of leaving their father's pouch. Their diet changes as they grow. Typically they will start out by eating tiny copepods, ostracods, and foraminifera, but as their snouts increase in diameter, and their energetic requirements increase with increasing size, they shift to a higher proportion of crab larvae (Brachyura) and mysid shrimps. In captivity, brine shrimp (*Artemia*) are commonly fed to seahorses, but lack adequate nutritional value. Much work has been done recently to improve seahorse husbandry techniques and determine appropriate nutritional supplements for captive seahorses.

PREDATORS

Seahorses' external armor makes them a fairly unappetizing meal for most potential predators, and their camouflage hides them effectively. That said, 82 predators of seahorses are known, including fish, octopuses, and birds. Seahorses have been found in the stomachs of several types of fish, including dolphinfish, dorado, jacks, tuna, snappers, and mackerel. Since these are generally pelagic fish, it may be that they are feeding on seahorses that have been torn from their home holdfast, or are in floating algae or debris. Seahorses may be an incidental food for many species, but it is unlikely that any species depends entirely on seahorses for their sustenance since seahorses tend to live at such low densities, and their bony bodies probably reduce their nutritional value. Seahorses with missing tails or fins attest to the fact that some benthic-dwelling creatures, such as crabs, also attack and may eat seahorses.

The most dangerous predators of seahorses, however, are humans. Throughout the world, but particularly in Asia, seahorses are prized for their medicinal or tonic qualities. They are used extensively in traditional Chinese medicine (TCM), but also in other traditional medicines including Indonesian *Jamu* and Japanese *Kampo* (see Trade, p. 48). They are targeted in some parts of the world (for example, in the central Philippines) but in many other areas they are simply retained from among the by-catch of trawl fisheries (see Destructive fisheries, p. 58).

ABOVE Surprisingly even birds take seahorses from their shallow water habitats, or if they find them floating or washed up on the beach. Seahorses have been found in birds' nests and even in the stomachs of Little Penguins.

OVERLEAF A watchful *H. reidi* keeps an eye on a floating crustacean, ready to snap it up for dinner.

COURTSHIP AND REPRODUCTION

PAIR BONDS AND GREETINGS

An endearing feature of seahorses is the way they form faithful pair bonds, which is relatively unusual in nature. Most seahorse species that have been studied in the wild show elaborate social interactions that create and maintain these partnerships. Every day, at least during the breeding season, paired male and female seahorses come together for a few minutes to greet one another. During this greeting they shed their typical concern for hiding and camouflage and visibly brighten in color, twirl in a complex dance around one another, and often entwine tails and swim in synchrony. When the time comes for the seahorses to mate, their daily greeting is essentially prolonged into a full courtship display (see Courtship and mating, p. 31). It has been shown, however, that it is regular greetings, rather than the act of mating, that solidify the pair bond.

It makes sense for species such as seahorses, which live at extremely low densities and do not move around much, to form long-lasting pair bonds. This way they can maximize their reproductive output (i.e. the total number of young they can produce) and avoid wasting time looking for a mate. The daily greetings also serve as a way for the male and female seahorse to match their reproductive cycles, again increasing their reproductive output.

Unfortunately the pair-bonding behavior of seahorses also has its drawbacks. In particular, it increases their vulnerability to disturbance. For example, if one member of a pair is caught in a trawl, swept away, or dies, the surviving seahorse may not find another partner and may therefore lose its reproductive potential. It has been shown experimentally that if a pregnant male *H. whitei* is removed, his female partner will not mate during the span of his pregnancy, even if another potential male is available.

Some of the species that have been shown to exhibit pair-bonding behavior include *H. comes*, *H. erectus*, *H. fuscus*, *H. guttulatus*, *H. hippocampus*, *H. kuda*, *H. reidi*, *H. spinosissimus*, *H. subelongatus*, *H. whitei*, and the pygmy seahorses *H. bargibanti* and *H. denise*. Pair-bonding behavior has not been seen in *H. abdominalis*, which lives at higher densities than many seahorse species, nor in *H. capensis*.

COURTSHIP AND MATING

Courtship is generally an extension of seahorses' daily greeting, or may be an interaction without any previous warm-up. The male usually initiates by indicating his readiness to receive eggs. He vigorously pumps water in and out of his pouch, displaying both its size and its emptiness. If the female is also ready, she will join him in a mating dance. This generally involves color changes, intensification of contrast between pale background body color and dark spots or stripes (or a dark head in the case of *H. guttulatus*), fluttering the fins, swimming in parallel, and pointing their heads up toward the surface. Ultimately the two seahorses rise together in the water column, and bring their genitals together, the female extending her tube-like ovipositor into the male's slit-like pouch opening. Sometimes mating seahorses will have to rise several times in order to line up correctly, and occasionally batches of eggs get dropped and cannot be retrieved.

Once the eggs have been transferred, the male shakes back and forth to settle them into his brood pouch, the mating colors fade, and he retreats to a quiet place to begin his pregnancy. The daily greetings continue in those species in which they occur, and the female starts work on a new batch of eggs. It will take her more or less the same time to mature a batch of eggs as it takes the male to complete his pregnancy, ensuring reproductive synchrony between the partners. For pairs that remain together, the

OVERLEAF Two Long-snouted seahorses (*H. guttulatus*) ready for mating. The male, on the right, shows his empty pouch to the female.

ABOVE Two Big-bellied seahorses (*H. abdominalis*) copulating.

synchrony of their reproductive cycles is often so precise that they will mate again within a few hours of the new father giving birth.

The males are technically the "limiting sex" (meaning that it is the rate at which a male can produce young that limits the reproductive rate of the population as a whole). However, there is no apparent "sex-role reversal" in mate competition among seahorses (i.e. males still compete for access to females, as they do in most species where mate competition occurs, rather than vice-versa, as is the case in some pipefishes where the females are more colorful, competitive, and active in courtship).

Pair-bonded seahorses tend to be monogamous, not only socially but also genetically, meaning that they mate with only one partner, at least within a single breeding season. This has been demonstrated for many species (including *H. comes*, *H. guttulatus*, *H. hippocampus*, and *H. zosterae*) based on field studies, and genetic analyses. In *H. breviceps*, if the first female does not seem to be receptive, males may display to at least two others. However, evidence to date suggests that the females maintain monogamy in the species by only choosing to reciprocate with the male they are partnered with, despite these extra-pair advances. Polyandry is known from *H. subelongatus*, with females occasionally switching partners between mating episodes, even when the previous partner was still available. Simultaneous polyandry has been shown in one case in *H. denise*, where a single female maintained two pair bonds over a period of time, mating with both males. The cost of this behavior was that neither male received a particularly large clutch of eggs. However, it allowed the female to spread the risk of losing her partner and hence her offspring.

After mating the pouch is sealed, precluding the deposition of additional eggs by another partner. This differs from the situation in the pipefish *Syngnathus typhle*, which has a "zipper-like" closure to the pouch. Males of this species can receive eggs from multiple females, and brood them simultaneously.

The breeding season for seahorses depends on the species. Temperate seahorses tend to breed only in the warm summer months, whereas tropical species and the large *H. abdominalis*, which occurs in the temperate waters of Australia and New Zealand, are fertile year-round. Seahorses in captivity can also be induced to breed year-round by manipulation of ambient temperature and photoperiod.

OVERLEAF This mating pair of *H. bargibanti* come together, while still attached to their home seafan, to transfer eggs.

EGGS, PREGNANCY, AND BIRTH

Seahorses produce relatively large pear-shaped eggs that develop continuously from spiral sheet-like ovaries. The unusual egg shape increases their surface area by 9 percent, compared to a spherical egg of identical volume, and so presumably increases their oxygenation within the males' pouch. The eggs are orange in color as a result of the high quantities of carotenoid-rich oil they contain. Seahorses are unable to synthesize their own carotenoids, so these essential pigments are derived from their crustacean diets.

After the female deposits her ripe eggs into the male's pouch, and the male's sperm fertilizes them, the pouch wall undergoes a dramatic change. It becomes thick and spongy, with extra layers to increase its surface area, and more blood vessels to serve it. What follows is physiologically complicated, and can be considered a true pregnancy. The fragile eggshells break down, and the embryos (each within its own embryonic sac) become embedded in the spongy walls. Here they begin to develop, nourished primarily by their attached yolk sac, but also receiving oxygen and some nutrition from the male. Over the course of the pregnancy waste materials are removed, and the fluid within the pouch becomes progressively more seawater-like, presumably easing the transition for the baby seahorses when they are released into the outside world. The hormone prolactin mediates these changes.

The number of offspring depends on the species. On average, seahorses produce between 100–300 young per brood, although the extremes range from 10 in *H. zosterae* to more than 1,500 in *H. reidi.* Typically not all eggs within a brood survive, and up to a third may perish within the pouch. In *Syngnathus typhle* it has been shown that the nutrients that these young contain are reabsorbed by the father and not taken up by other developing embryos. However, where these extra nutrients end up is unknown for seahorses.

Seahorse pregnancy usually lasts from between two to four weeks, depending on the species and the water temperature. It is about 12 days for *H. zosterae*, 14 days for *H. comes*, 20–22 days for *H. whitei*, and 30 days for *H. abdominalis.* It is energetically costly for the male. For example, the metabolic rate in brooding *H. zosterae* has been shown to increase by 10–52 percent, in comparison to non-brooding males. This extra energetic expenditure could be one reason why male seahorses tend to maintain smaller home ranges than do females.

Birth is a seemingly painful process that may take as little as ten minutes, or may last for up to three days. The male undergoes strong contractions, initiated (as in humans) by the hormone oxytocin, and jack-knifes his tail against his pouch, in order to expel the young. Sometimes he may use a holdfast, the ground, or a convenient rock or shell to help press against the pouch and push the young out.

There may be a "warning" birth of one or a few young a day or two before the main birth event, but once the young are out of the pouch they must fend for themselves.

In the BBC documentary *Kingdom of the Seahorse* (1996), a pregnant male *H. whitei* was brought to the gynecology department of one of the Sydney hospitals and his pouch young were filmed using a hysteroscope (a device normally used to examine a woman's uterus). It was undoubtedly the only time that the department had a pregnant *male*, rather than female, on the examination table.

BELOW The skin of the previously distended pouch looks flabby after a male has given birth to his brood. In a bonded pair a male will not have to wait long before he receives another batch of eggs from his partner.

SEAHORSE OFFSPRING

Baby seahorses are born looking like miniature versions of their parents. They have a grasping tail, which is sometimes used to attach to siblings, forming a writhing ball of newborns. They also have fully functional dorsal and pectoral fins for locomotion. Some species, such as *H. kuda*, *H. mohnikei,* and *H. spinosissimus* even have a vestigial caudal (tail) fin that is lost in adulthood. Young seahorses tend to be straighter (more like pipefishes or pygmy pipehorses) than adults. They also have larger heads and eyes in relation to their bodies, and shorter tails. Some are born with the yolk sac still attached and can use that for nutrition for the first day or two after birth, however, in most newborns the yolk sac is fully absorbed prior to birth and the baby seahorses must start feeding immediately. Juvenile seahorses have shorter, deeper snouts than adults, and it turns out that this is actually ideal for the low Reynolds number environment (i.e. extremely viscous in comparison to their size) in which they live. A short, deep snout maximizes

suction, and it is this, rather than high velocity of head rotation, that is required in order to set the relatively viscous fluid in motion.

In captivity, newborn seahorses tend to rise to the surface of the water, although less so today with modern culture techniques. This can be a dangerous proposition for such a small creature, as they may get stuck in the surface tension and be unable to descend again. They may also end up "over-inflating" their swim bladder, causing premature death. However, those that succeed in adequately inflating their swim bladder immediately set about eating.

No-one knows exactly how long seahorses remain in the water column, and hence have the opportunity for dispersal, before they settle to the sea floor to live out their adult lives. Independent from birth, baby seahorses are pigmented and vulnerable. At first their tails may not be long enough to adequately attach to a holdfast, but they grow quickly. Aquarium-based developmental studies suggest that there are three growth stages, the most rapid being the second, which likely coincides with the shift from pelagic to benthic life. During this time there is an abrupt increase in tail growth, after which they start grabbing at holdfasts (including each other, if nothing else is available) and swimming activities are reduced.

Young seahorses, and even occasionally adults, have been found in plankton tows, but they are not common. Usually they are found near the surface, but some specimens have been brought up from depths of more than 195 feet (60 m). It is believed that most species stay in the plankton for a maximum of two to four weeks. Compared with other reef fishes, which commonly have pelagic larval durations of 30–120 days, this is a relatively short time during which to disperse. By the time the seahorses settle, they have reached about 3/4 inch (2 cm) in height, depending on the species.

Pelagic young tend to be more elongated in their body shape than bottom-dwelling adults. In fact young *H. abdominalis* were originally described as a different species (*H. graciliformis*) because they looked so different. The same happened to young *H. ingens*, erroneously described as *H. hildebrandi*.

It is possible that some individuals may remain planktonic later in life. In the winter of 2006, hundreds of seahorses washed up on beaches in South Australia. These seahorses were adult males and females and it is unclear why they all suddenly came ashore. It is also not clear exactly which species they should be assigned to, as their morphology did not completely match either of the two local species, *H. abdominalis* or *H. breviceps*. The Inshore Fish Group has been trying to gather more information to help solve this mystery.

OVERLEAF Newborn baby seahorses have large heads and eyes in relation to their bodies, but are otherwise miniature versions of

DISTRIBUTION

HABITATS

Seahorses are found in all tropical, subtropical, and temperate seas, although their distribution is patchy and they tend to live at very low population densities. They can be found in a range of habitats, including seagrass, coral reefs, mangroves, sponge gardens, and even on bare sand. Generally they require a holdfast to grab onto with their tail (see p. 25). Some species use different habitats at different stages of their lives. For example, juveniles of the coral-dwelling *H. comes* are commonly found in *Sargassum* beds.

Species that live in the same geographical area may separate themselves based on habitat preference. For example, on the Mediterranean and Atlantic coasts of Europe *H. hippocampus* prefers more open habitats, whereas *H. guttulatus* prefers vegetated areas, and in the Philippines, *H. comes* is typically found on coral reefs, whereas *H. kuda* is found in estuaries and mangrove areas.

Most seahorses live in shallow water, from the intertidal zone to about 66 feet (20 m). Some, such as *H. erectus* (in their northern range), *H. guttulatus,* and *H. hippocampus* undergo seasonal migrations to offshore, deeper waters in the colder month. Others may use deeper water during their dispersal. *H. spinosissimus* and *H. trimaculatus* are commonly brought up by trawl fishermen in Southeast Asia from depths of about 49–131 feet (15–40 m) on sandy bottoms. And, *H. kelloggi* are rarely brought up from depths of less than 66 feet (20 m). *Hippocampus minotaur* and *H. paradoxus* are deep-water species known only from the type specimens that were collected from between 197 and 361 feet (60–110 m). A specimen of *H. patagonicus* was captured 75 miles (120 km) offshore at a depth of 197 feet (60 m) during a survey for hake, and there have been reports of specimens collected from similar depths off the east coast of Africa. *H. ingens* is regularly recorded at depths below 131 feet (40 m) around the Galapagos Islands. A seahorse from the Northwest Hawaiian Islands was recently found by a team of divers using rebreather apparatus and diving at depths of up to 285–288 feet (87–88 m). This is one of the deepest seahorse records, and certainly the deepest that has been seen alive (and photographed) *in situ.*

If we know where seahorses live we can do more to protect them (see Conservation, p. 55). Habitat protection is paramount, particularly limiting destructive trawling and blast fishing, and educating divers to avoid damaging and breaking corals and gorgonians while underwater. Removing unwanted fishing gear, and cleaning or replacing underwater man-made objects, although generally a good idea, can sometimes be detrimental to local seahorses populations. In Australia, seahorse biologist David Harasti, has developed a best-practice procedure for cleaning protective swimming nets in Sydney Harbour. These nets become rapidly overgrown with bryozoans, ascidians, algae, and other epibiotic growths, the weight of which can severely damage the net. The nets are, however, also the home of large numbers of seahorses (*H. abdominalis* and *H. whitei*). The best practices put forward by Harasti include leaving the growth within 4 feet (1.2) m of the ground where the greatest densities of seahorses exist, cleaning the nets one small section at a time, and cleaning in the winter months when fewer seahorses use the nets, and when the seahorses are not breeding.

DISTRIBUTION MAP

BELOW Seahorses are distributed throughout the world, in both tropical and temperate zones, ranging from about 45 degrees north to 45 degrees south

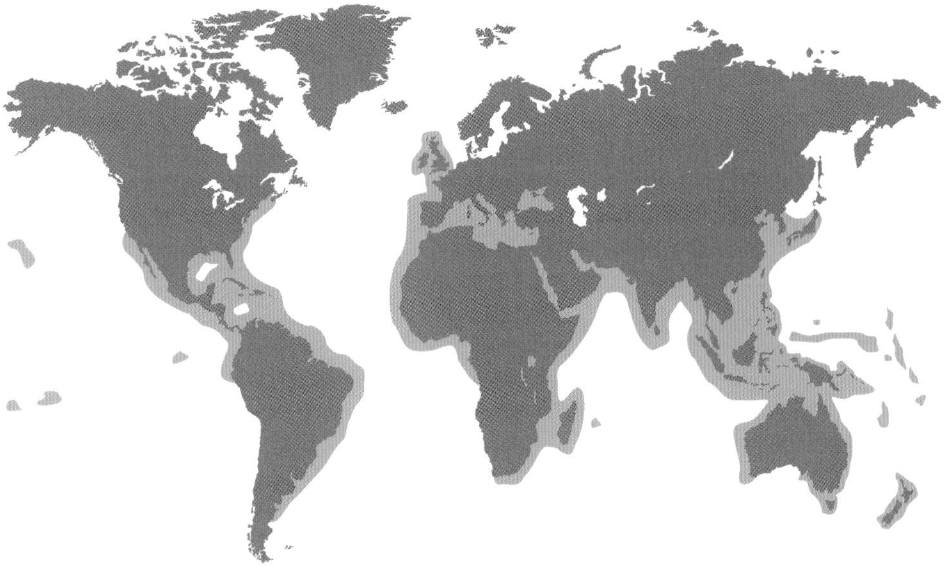

DISPERSAL

The main period during which seahorses disperse is probably when they are pelagic (floating) larvae. This period is believed to last for about two to four weeks, and is one of the least understood parts of seahorses' lives. Relatively few seahorses are caught in plankton tows (surveys of the floating animals and plants that drift with the ocean currents), yet millions of adults are caught in trawl nets.

Despite having a relatively short pelagic period, seahorses have somehow spread across the globe. It is believed that rafting may have played an important role in this. Young that attach to drifting algae or debris can travel far, and if adults also use this mode of transport it would only need a single pregnant male to start a new population.

Genetic studies support the hypothesis of rafting, and rare founder events initiating entire populations. For example, within the *H. kuda* group (clade) there are multiple examples, from all across the Indo-Pacific, of populations that consist primarily of a single, common, ancestral genetic type (haplotype), which may have come from a single founder, and a small number of other haplotypes which are extremely closely related, that may have evolved since the population was founded. The patchy distribution, with no clear evidence of "isolation-by-distance" (where populations close to each other are more closely related to each other than those further apart) among many seahorse populations also supports the idea of relatively rare, almost random, founding events across large geographic distances.

Other recent reports also support the passive dispersal/rafting hypothesis, for example: a specimen of *H. erectus* (a western Atlantic species) was found in the Azores (off the coast of Africa); juveniles of *H. patagonicus* were captured in open water attached to floating debris; and mass strandings of young seahorses occurred in South Australia, along with some (but not huge amounts) of algae.

MOLECULAR EVIDENCE

Molecular data, such as DNA sequencing, microsatellites, or restriction fragment variation, have great potential to provide information about the history, size, and inter-relatedness of seahorse populations. These in turn can help disentangle theories about life history and dispersal, as well as help with assessing species limits, species identification and conservation planning.

The interplay between molecular variation and geography is called phylogeography and is influenced by a species' life history, behavior, and external factors that have operated on it over time and space. A comparison of the phylogeographic patterns for four species in Southeast Asia showed some interesting results. The two shallow water species (*H. barbouri* and *H. kuda*) had much more fragmented populations than did the two deeper water species (*H. spinosissimus* and *H. trimaculatus*) for which shared genetic types (or haplotypes) could be found scattered throughout their distribution. In the case of *H. trimaculatus* the same haplotype could even be found at the opposite extremes of its range, in Japan and India. It is postulated that the fragmented structure of shallow-water species reflects the distribution of their preferred habitats, which are separated from one another by wide areas of uninhabitable deep water, as well as effects of ocean basin isolation during periods of low sea level during the most recent ice ages. By contrast, the habitat for the deeper-water species may be more continuous, at least along the continental shelves.

H. trimaculatus

SHALLOW-WATER SPECIES

LEFT AND BELOW *H. kuda* (left) and *H. barbouri* (below) are two shallow-water species that show genetic isolation among populations across their range.

DEEPER-WATER SPECIES

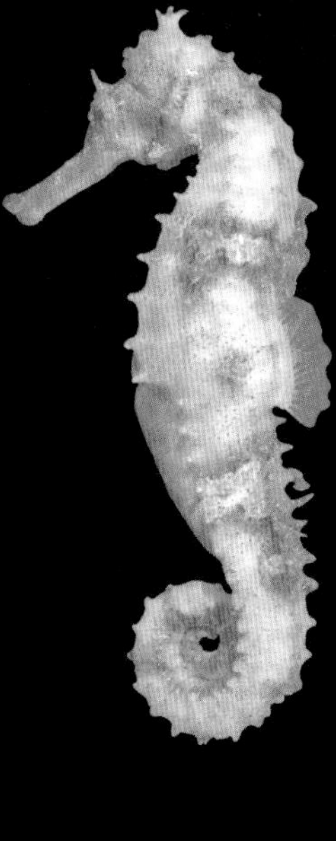

LEFT *H. trimaculatus* (far left) and *H. spinosissimus* (left) are deeper-water species that show less variation in their DNA sequences across the Indo-Pacific.

FOSSIL SEAHORSES

FOSSIL EVIDENCE

It is only relatively recently that fossils of seahorses and their close relatives have been found, despite having a bony exterior that would seem to predispose them to excellent fossil preservation. In 1978, more than 2,000 fish specimens (including 37 families and 48 genera) were found at a site in eastern-central Italy (Marecchia River) that dated from the Pliocene, approximately 3.1 million years ago (mya). These included several seahorse specimens that were identified as the Long-snouted Seahorse (*H. ramulosus* = *H. guttulatus*), as well as specimens of the pipefish *Syngnathus acus* and a large unidentified syngnathid (> 15¾ inches/40 cm long).

The oldest known seahorse fossils were found in laminated siltstone rocks in a stream in the Tunjice Hills of Slovenia, in 2007, by paleontologists Jure Žalohar and Tomaž Hitij, who were looking for insect fossils for their research. At the time that these seahorses are believed to have been alive, 12.5 mya, this area would have been the shore of a huge intra-continental sea, the Paratethys (the remnants of the global equatorial seaway called the Tethys). The fossils were mostly juveniles and have been named *H. slovenicus* (after Slovenia, the country they were found) and *H. sarmaticus* (referring to the time period from which they originate). Both species are fossil taxa only and have no living representatives.

The oldest syngnathid fossils are *Prosolenostomous lessenii* from Monte Bolca in northern Italy (Eocene, 48–50 mya). More recent fossils of seahorse relatives include members of the genera *Doryrhamphus* (from the Caucasus), *Hipposyngnathus* (Carpathians, Poland, Caucasus, California, and Italy), *Nerophis* (Austria), and *Syngnathus* (Caucasus, Poland, Italy, and California). Very recently (2012), a fossil pipehorse (*Hippotropiscis frenki*) was found in the same area of Slovenia as *H. sarmaticus* and *H. slovenicus*, also dating from about 12.5 mya.

Fossil evidence suggests that seahorses may have evolved from an ancestor that had a fused, slightly elongated snout, and a more elongated body than modern-day seahorses. This ancestor probably swam horizontally, as do pipefishes and pipehorses. One of the main evolutionary innovations of seahorses was the adoption of their upright posture. This new arrangement necessitated some significant morphological changes, particularly in the spine, in order to keep the head horizontal. A vertical posture, and longer snout, would also have increased the efficiency of seahorses' pivot feeding (see Adaptations of the head, p. 13), and may have contributed to the evolutionary pressures that shaped the current seahorse body plan.

It is believed that seahorses diverged from the pygmy pipehorses approximately 28 mya, during the Oligocene period. It was around this time that significant tectonic changes were taking place and huge expanses of shallow sea existed (particularly between Australia and Asia). These conditions were ideal for the development of vast seagrass beds, which in turn, it has been proposed, provided a perfect habitat for the origin of the upright seahorse. The vertical posture, it is hypothesized, gave the seahorse a selective advantage by helping to conceal it from predators among the vertical blades of seagrass. Since some of the most basal seahorse species (*H. abdominalis* and *H. breviceps*) are found only in Australasia, and Australia is also the home to the three species of *Idiotropsiscis* (the most closely related pygmy pipehorse species), it is believed that Australia may have been the original birthplace of the seahorse genus, possibly in the extensive seagrass beds that extended between northern Australia and Indonesia at that time. The more horizontal swimming position of *H. abdominalis* (especially juveniles) suggests that it is a relatively primitive species within the genus.

Despite earlier claims (and its name), it is unlikely that *Hipposyngnathus* was an ancestor on the direct line to seahorses because its pouch is positioned on the trunk, not on the tail (i.e. it is part of the Gastrophori, and not the Urophori lineage, see page 9). *Hippotropiscis,* on the other hand, has its brood structure on the tail, and is probably more closely related to seahorses.

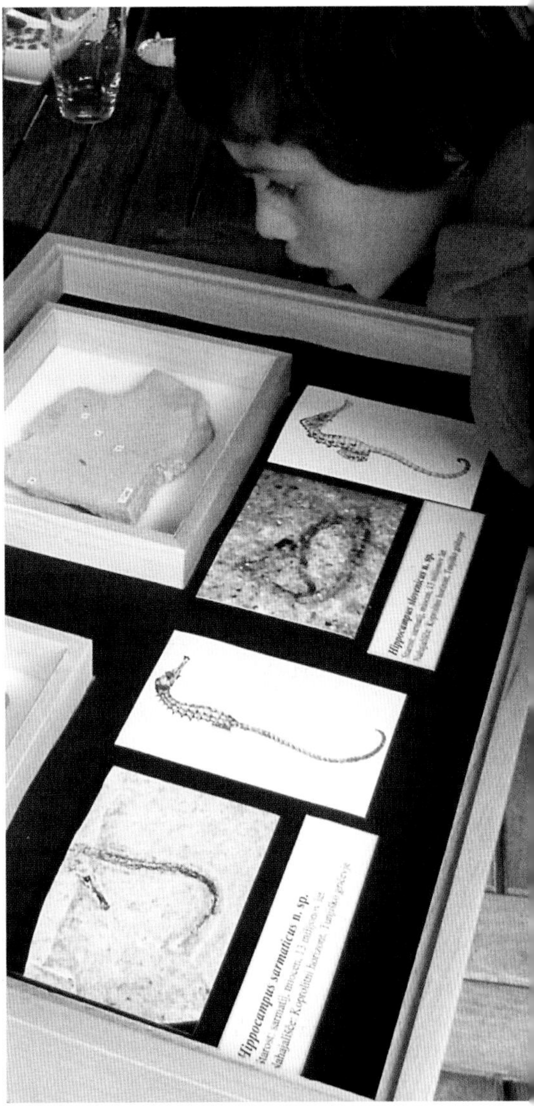

RIGHT The author's son, Oliver Laxer, examining 12.5-million-year-old seahorse fossils in Slovenia.

EVOLUTION

GENETICS AND PHYLOGENY

Species are groups of individuals that can mate and produce viable offspring with one another, but do not normally breed with members of another such group. For practical identification of individual seahorses, morphological and/or genetic characters are usually used as a proxy to determine species membership. Specimens that share what are called "homologous characters" (i.e. features that have evolved in the same way) are deduced to be related, and if they share all of their features, to be in the same species. A hierarchy of relatedness creates a hierarchy of taxonomic levels from species (smallest scale), to genus, family, order, and so on.

Relationships among seahorse species, and among seahorses and other members of the Syngnathidae family have been elucidated not only from fossils and morphology (for example, counting tail rings, analysis of bones, and pouch development), but also increasingly using genetic evidence. Allozymes (variant versions of enzymes that have different electric charges) can be used to determine genetic similarities among specimens. So, too, can DNA sequencing. This involves determining the actual sequence of DNA code—the sequence of the four component nucleotides: adenine (A), guanine (G), cytosine (C), and thiamine (T) within genes of different specimens. In recent years, as sequencing methods have become more standard, and costs more affordable, there has been a huge increase in the availability of DNA sequence data and as a result we have learned much about the species' relationships to one another.

MITOCHONDRIAL DNA

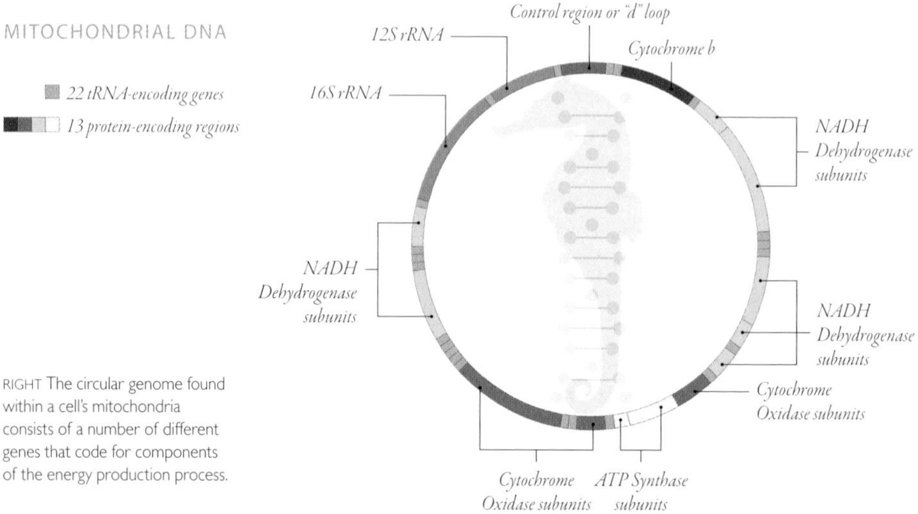

RIGHT The circular genome found within a cell's mitochondria consists of a number of different genes that code for components of the energy production process.

Today there are more than 2,000 *Hippocampus* sequences available on GENBANK, an online repository of sequence data. Most of these are from the mitochondrial genome (the DNA that resides within the part of the cell responsible for producing energy) but many are also from the nuclear genome (the DNA that resides within the nucleus of the cell). Mitochondrial DNA is inherited only through the mother (as mitochondria are passed down the generations through the egg), whereas nuclear DNA is inherited through both parents.

Comparing the sequences across specimens allows researchers to build phylogenetic trees (like family trees) based on similarity. These represent hypotheses of evolutionary relatedness among the species. The closer two branches are in the tree, the more closely related are those two species. The original genetic work on seahorses (carried out by Steve Casey at the Zoological Society of London) involved sequencing, for 28 different species, the section of the mitochondrial genome (the gene) that codes for cytochrome *b*, a protein that is involved in electron transport in the respiratory complex. Since then other researchers have looked at other gene regions including the mitochondrial control region and cytochrome oxidase 1 gene. More species have been included, and other studies have considered the relationships among populations within species.

Cytochrome oxidase 1 is the gene of choice for the Barcode of Life (BOLD) project and many of the species held at the Redpath Museum, McGill University, Canada, have been sequenced and included in the BOLD reference database, providing interesting information about the relationships among seahorse species. Genetic distances between species can tell us how closely related they are to one another. The exact distances depend on the time since the two species began to diverge from one another, and the speed with which a particular gene fragment evolves. However, to give a general idea, based on the mitochondrial cytochrome oxidase 1 gene, genetic distances within a seahorse species are usually < 2 percent whereas those between species are usually 7–23 percent. The results do show that a number of recognized species (particularly in the *H. kuda* clade) are genetically very close to one another and may not ultimately prove to be separate species. At the other end of the spectrum, there are specimens that have been morphologically identified as the same species (for example *H. histrix*) from Southeast Asia and Mozambique that differ by as much as (or more than), many other "good" species, suggesting that they may in fact represent more than one species. Clearly more work is needed in order to sort out the taxonomy (the science of defining and naming species) of seahorses, and this is the subject of ongoing research.

MAJOR CLADES WITHIN THE GENUS *HIPPOCAMPUS*

Although seahorses display such a bizarre mixture of physical features—grasping tail like a monkey, pouch like a kangaroo, independently moving eyes like a chameleon, and a suit of armor like a lobster—they are remarkably conservative in their form across the genus. Most species have 11 trunk rings, between 34 and 40 tail rings, and between 15 and 19 dorsal and pectoral fin rays. The most morphologically distinct species are *H. abdominalis* with 12 or 13 trunk rings, 45–48 tail rings, and 25–29 dorsal fin rays, and *H. minotaur* with eight trunk rings, 41 tail rings, and seven and 11 dorsal and pectoral fin rays respectively. This lack of variation in morphological characters has contributed to the confusion in seahorse taxonomy and may reflect some kind of constraint on their evolution. It seems that the general seahorse body plan, once settled upon, remained pretty constant as the species diverged. Presumably it was a design that worked well, so there was no need to change it. That said, there are definite groupings (or clades) of seahorses that are more closely related to one another than they are to other species, and this has been further elucidated by genetic data.

Work done by Steve Casey, Peter Teske, Tony Wilson, Healy Hamilton, myself, and others has confirmed that seahorses form a monophyletic group (i.e. all seahorses derive from a single common ancestor) and are more closely related to one another than they are to any non-seahorse relative. It has also shown that within the genus, there are several subgroups. One of these is a circumglobally distributed group of relatively smooth seahorses. This has been termed the "*kuda*-clade", and it includes *H. kuda*

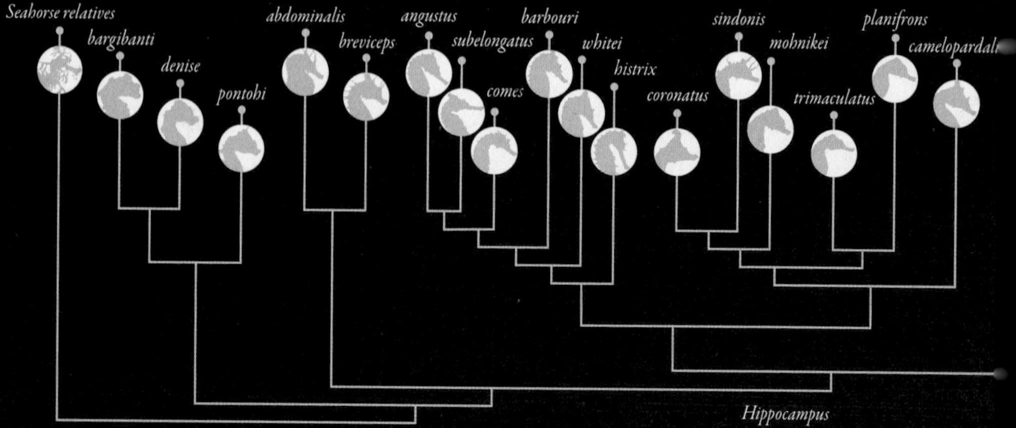

as well as *H. algiricus*, *H. capensis*, *H. fuscus*, and *H. reidi*. Another group is the "*H. histrix* clade", which includes many Southeast Asian and Australian species. These are often spiny, with a double spine on either side of their throat, a striped snout, and a pattern of fine brown reticulated lines on their bodies. This group includes *H. angustus*, *H. barbouri*, *H. comes*, *H. subelongatus*, and *H. whitei*, as well as a number of other seahorses from northern Australia. Research is yet to establish whether the latter represent a single, physically variable species, or multiple species each with a very restricted geographic distribution. In the Atlantic we find the "*erectus*-clade" that spans the ocean, with *H. erectus* and *H. patagonicus* off the coasts of the Americas and, sitting in between them on the evolutionary tree, *H. hippocampus* in the eastern Atlantic and the Mediterranean. The pygmy seahorses, *H. bargibanti*, *H. colemani*, *H. denise*, *H. pontohi*, and *H. satomiae*, apparently diverged from the rest of the seahorses early on in the evolution of the genus. They form a group by themselves, at the base of the seahorse tree. Also near the base of the tree is a group comprising two southern Australian species, *H. abdominalis* (also in New Zealand) and *H. breviceps*. The genetic relationships of a few species (for example *H. guttulatus* and *H. zosterae*) are uncertain, and some species have not yet been sequenced (for example *H. debelius* and *H. paradoxus*). In the Species section of this book, we group the species based on their evolutionary relationships, to the extent that this is currently known.

BELOW A tree of evolutionary relationships among seahorse species based on DNA sequence data. Not all species have been studied genetically, so some species are missing from this figure. Portrayed distance on the figure does not necessarily imply level of divergence among species.

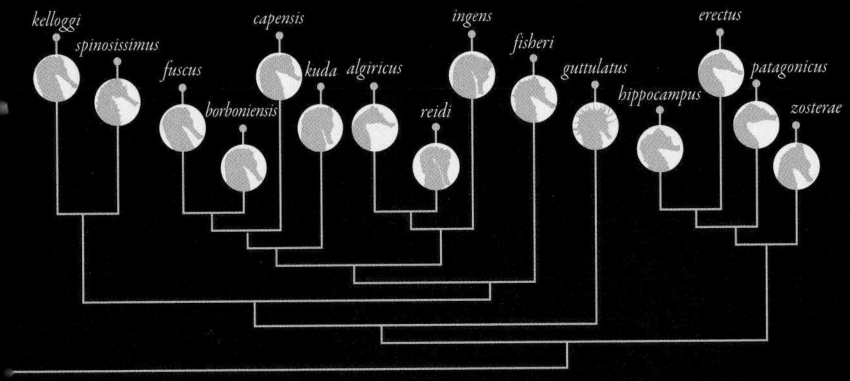

GENUS GROUPS AND SUBGROUPS

TRADE

TRADITIONAL MEDICINE

Seahorses have been used medicinally for hundreds, if not thousands of years. Traditional Chinese medicine (TCM) considers them a key ingredient to treat many disorders. This form of medicine has been codified since the 3rd century BCE, but began far earlier. The first definitive reference to seahorses in TCM, however, comes from Chen Cang-Qi's herbal (Ben Cao Shi Yi), in AD 739.

The philosophy of traditional Chinese medicine is somewhat different from that of western medicine, which focuses on specific ailments and has targeted cures. In TCM the whole body is treated as an intricately balanced system of *yin* and *yang*. Diseases and pains are symptoms of imbalance within the body, and restoring the balance recreates harmony and health. Seahorses (*hai ma*) are considered to be strong *yang* tonics, sweet and warm, and in combination with other ingredients they help strengthen the kidneys (*yin* organs) and blood.

Seahorses are among the most valuable TCM ingredients by weight. Large, pale, smooth seahorses are the ones most desired by TCM practitioners, and so seahorses are often bleached, once dried, in order to attain a pale color. Typical methods for taking seahorses medicinally include submerging them whole in alcohol (as 'seahorse wine'), grinding and mixing them with other ingredients, as a decoction, or in powdered form. Recently there has been a move toward the development of pre-packaged TCM medicines. This opens the door to the exploitation of all seahorses, however small, dark, spiny, or otherwise unattractive they may be to discriminating end users. This trend is a serious conservation concern, opening the door

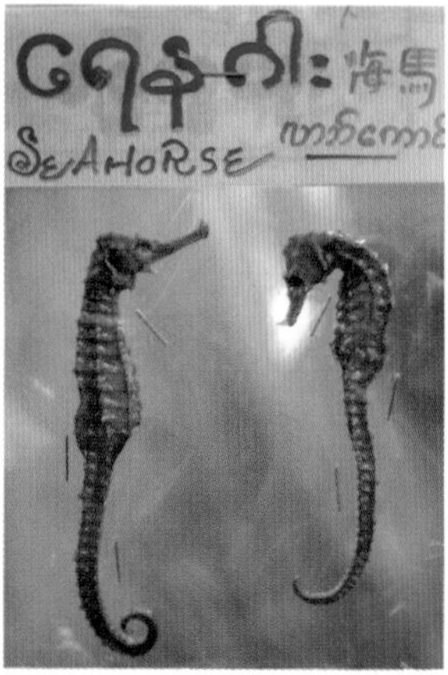

FAR LEFT Dried seahorses (and pipefishes above) for sale in a Hong Kong market.

LEFT Dried seahorses for sale at the night time tourist market along Chaungtha Beach, Myanmar. The beach is popular with foreign tourists and weekenders from Yangon.

for trade in all species, and all age classes. Also, if pre-packaged medicines are prepared in source countries, it makes it very hard for monitoring and enforcement agents to check the accuracy of permits in terms of species, volumes, and so on.

In the west, seahorses appear in the Greek physician, pharmacologist, and herbalist Pedanius Dioscorides' five volumes of "*De Materia Medica*", published between AD 50–70, as a cure for baldness. They have also been recommended, by numerous texts dealing with TCM, to cure respiratory ailments, problems with kidneys, and abdominal pain, as well as to slow the aging process, cure arthritis, and to aid childbirth and sexual function. Recent research, into the mechanisms behind the action of seahorse extracts, suggests that they do have antimicrobial properties, produce peptides that prevent the breakdown of chondrocyte cells (those that create and maintain healthy cartilage), and have free radical-scavenging abilities.

Although seahorses are no longer typically used in western medicine, the demand for seahorses worldwide is great. At least 80 countries are involved in capturing, trading and consuming them. Experts estimate that 15–20 million seahorses enter the trade each year; this number is based on official statistics and on-the-ground research. Many more than this are caught, most of them accidentally, in non-selective fishing gear (see Destructive fishing, p. 58). The vast majority of the seahorses are exported dried, and mostly for TCM—about 90 percent of exports are destined for consumers in Hong Kong SAR, Taiwan, and China. Elsewhere in Asia, other forms of traditional medicine (for example Korean *Hanyak*, Indonesian *Jamu,* and Japanese *Kampo*) also use seahorses.

In northern/northeastern Brazil seahorses are one of the most popular animal items to be used in the Afro-Brazilian religion Candomblé. The magico-religious properties of seahorses are believed to include the ability to ward off the evil eye, eliminate harassing spirits, bring good luck, money, or business success, act as an aphrodisiac, and treat ailments including asthma, stroke, inflammation, sinusitis, alcoholism, and cancer. They are worshipped alive, may be ground to a powder, taken in *águas* (seahorse water), or worn as part of an amulet (*patuas*).

It is clear that humans worldwide have deep spiritual and cultural connections with seahorses. Belief in their healing properties is tied not only to their physical composition, but also to these more mystical components. As such, any dialog that advocates changes to human behavior in this realm needs to proceed in a way that is respectful of these different worldviews and involves all stakeholders. That such a shift may be possible is supported by the fact that some TCM practitioners in Hong Kong SAR, and those in western countries, at least, are open to sustainable sourcing of seahorses, and even using alternatives. For example, after a conservation workshop, a group of TCM traders from Hong Kong

SAR wrote to their members asking for a commitment to only using seahorses greater than 4 inches (10 cm) in height (ones that presumably would have reached sexual maturity prior to being captured, see figure on p. 57). They too want seahorses to be around in perpetuity, if not for intrinsic value, then for their value as a medicinal product.

AQUARIA

The practice of keeping fish in aquaria dates back to Roman times. However, it really took off as a hobby during the 19th century, particularly in England. Today there are nearly 10 million households, in the USA alone, that maintain aquaria (it ranks as the second most popular hobby after photography), of which at least 700,000 are saltwater aquaria. Most of the inhabitants of these glass-walled cages, including seahorses, are wild-caught, and 70–90 percent die within a year. Seahorses, in particular, are notoriously difficult to maintain in captivity. They are very picky eaters, and are prone to a variety of diseases, including developing gas bubbles under their skin, a range of bacterial infections, and tail rot.

Captive-bred seahorses tend to have longer life expectancies in aquaria than their wild-caught relatives. They have been brought up, and hence acclimatized to, eating frozen or dried foods, the typical diet provided in home aquaria. Normally seahorses eat only live animals. Captive-bred animals probably also survive the transportation to a new home more readily than wild-caught ones, which may have been removed from their faithful partner as well as their home turf, and spent many hours or days in small, airless plastic bags during transportation.

Public aquaria typically have more resources, space, and access to experienced aquarists and vets than do hobbyists. Many of these institutions display seahorses, and some create elaborate exhibitions focused on these remarkable animals. For example the John G. Shedd Aquarium in Chicago created an award-winning exhibition entitled "Seahorse Symphony" that opened in 1999 and ran for an unprecedented five years, and included a musical commission—*Seahorse Serenade*

by Augusta Read Thomas for the Chicago Symphony Orchestra. The exhibition was attended by more than 2 million visitors per year, and served as a fantastic opportunity for education, highlighting not only the diversity of seahorses and their relatives, but also the plight of seahorses worldwide, and the conservation work of Project Seahorse in the Philippines.

The Birch Aquarium in San Diego also developed and hosted a wonderful, ongoing seahorse exhibition. The aquarists at the Birch Aquarium started a Seahorse Propagation Program in 1994. Since then, the program has successfully bred 13 different species of seahorses, and shipped more than 3,000 individuals to more than 65 other public aquaria for display.

Other public aquaria have engaged hugely with seahorse displays, education, husbandry research, conservation, and funding (including Monterey Bay Aquarium, USA; National Marine Aquarium, UK; National Aquarium, Baltimore, USA; Moody Gardens, Galveston, USA; Ocean Park, Hong Kong SAR; Nausicaa, Boulogne, France *et al.*), arguably none more so than the Zoological Society of London (UK), a founding partner of Project Seahorse, and home to Project Seahorse Foundation in the Philippines.

ABOVE RIGHT Dried seahorses make unusual beach souvenirs.

OVERLEAF The John G. Shedd Aquarium in Chicago that hosted the "Seahorse Symphony" exhibit from 1999-2004.

CURIOS

The unusual shape of seahorses fascinates us, and the fact that they dry and preserve easily makes them ideal beach souvenirs. Whether they are sold simply as dried seahorses, or (dis) tastefully arranged among other dead sea life in plastic tombs, with fake snow fluttering about like bubbles in a make-believe plastic reef, or ignominiously spinning in yo-yos that advertise "Real Seahorse Inside," embedded in a toilet seat, garishly painted with their eyes punched out and turned into a key fob or magnet, or pasted onto a card to send to relatives who could not join the tourists on their beach vacation, they are still dead seahorses from the wild. The scale of the curio trade is global and extensive, and contributes to the conservation concern for these animals.

SOURCES

Where do all these seahorses come from? The majority (> 95 percent) comes from incidental capture by fishing gear (especially trawlers), which target bottom-dwelling species (for example, shrimp), particularly in seagrass and soft-bottom habitats. Seahorses are separated, and retained, from among the pile of by-catch (unwanted, non-target species) because fishers know that they have an economic value. A minority, but still significant quantity, comes from targeted fisheries. These include breath-hold divers (or hookah divers that have air pumped from a compressor on board a boat), who catch seahorses while also looking for other fish and sea cucumbers, shells, and other invertebrates. Some specimens come up attached to stationary fishing gear, such as crab pots, and a small, but increasing, number come from captive-breeding operations, primarily to supply the high-value aquarium market.

Amanda Vincent published the first report on the international trade in seahorses in 1996, and estimated that the global trade amounted to at least 19 million individuals per year. This report was based primarily on her research in Asia, and it was clear that this region was both the largest source, and the largest consumer, of seahorses. At that time, at least 32 different countries were known to be involved in the seahorse trade.

Since then, Amanda and her colleagues have undertaken additional trade surveys in other areas of the world, and official statistics have been collected that further our understanding of the extent of the trade. Based on data in the CITES (Convention on International Trade in Endangered Species) database, from 2004–2011, the seahorse trade apparently includes 87 Parties, from all continents except Antarctica. Of these, 61 were reported as exporting dried and live seahorses, 56 were reported as importing seahorses, and some were reported as doing both. These countries reported trading an average of 5.7 million individual seahorses per year (range 3.3–7.6 million), but given the evident gaps in reporting, lack of coverage of domestic consumption, and illegal, unregulated, or unreported (IUU) trade, this is likely to be a huge underestimate. Including other data sources, the total volume of the trade is estimated at between 15–20 million seahorses per year (if not substantially more), making it one of the largest wildlife trade issues in the world.

Based on the CITES data, Asia is still the primary source, with Thailand reportedly providing 88 percent of the dried specimens, followed by Guinea, China, Senegal, Malaysia, and Vietnam. Indonesia, Malaysia, and India have officially banned the export of wild-caught seahorses as a direct response to the CITES process. However, capture and trade does still continue, as it does in an ever-expanding number of nations, with West African countries being among the most recent to enter the trade. Fishing and export of seahorses from the Philippines was also banned, until a recent change to the fisheries law that has attempted to bring the national regulations more line with the spirit of CITES.

Again based on the same CITES data, the six most heavily traded species accounted for more than 90 percent of the trade. These were *H. trimaculatus* (31 percent), *H. spinosissimus* (30 percent), *H. kelloggi* (18 percent), *H. kuda* (6 percent), and *H. algiricus* (6 percent), and they mostly come from trawl fishing by-catch. The majority of these (> 98 percent) are dried and destined for the traditional medicine markets. For the live trade, the most common species are *H. kuda* and *H. reidi* (with a reported average of 65,000 and 29,000 traded per year respectively). The live-traded specimens tend to be from targeted fisheries, but there is also a growing captive breeding component (particularly from Sri Lanka). The top reported sources for the live trade were Vietnam, Sri Lanka, Indonesia, Australia, Brazil, and Mexico.

BELOW Seahorses can draw attention to the many millions of tons of unwanted by-catch that are brought up primarily by trawl boats operating in shallow seagrass and soft-bottom habitats.

The high price paid for seahorses ensures the continued involvement of fishers in direct capture of seahorses, for example in Vietnam, the Philippines, and Brazil, since seahorses can provide a significant proportion of a subsistence fisher's income. The by-catch issue, where again subsistence fishers benefit economically from retaining seahorses, is even more of a challenge in terms of conservation and legislation, especially since once seahorses are incidentally caught, they are often dead, damaged, or unlikely to survive even if returned to sea.

CITES

In response to conservation concerns, all seahorses were listed on Appendix II of the Convention on International Trade in Endangered Species (CITES) in 2002, and this came into force in 2004. Since then, all countries ("Parties") that are signatory to CITES are required to declare that their exports are "not detrimental to wild populations," provide export permits, collect data on imports, and submit annual reports to CITES of their international trade.

CITES documentation theoretically provides a quantitative measure of the trade in seahorses, and offers an unprecedented means to track and monitor the trade. In reality, problems with poor compliance, lack of accuracy in species identification, missing units in the reported volumes, and delays in processing the records, all contribute to limiting the usefulness of the data. CITES Parties also report only those seahorses that are exported internationally, and do not account for domestic consumption, which may be substantial in countries such as Vietnam and Indonesia. Critically, IUU trade is not taken into account, but could be extensive, particularly in Asian countries.

Several Parties have purportedly addressed the issue of having to prove sustainable exports under CITES by putting an end to official exports of wild seahorses (including India, the Philippines (until very recently), Indonesia, and Malaysia). Unfortunately, however, the large trade that existed prior to the implementation of CITES (all of these countries had been major suppliers) has not likely disappeared overnight, and may well have gone underground, making tracking and monitoring it even harder.

Implementing the CITES listing, and ensuring that it helps with conservation and sustainable trade, will require some work and dedication on the part of the Parties. However, it has brought the issue of sustainable trade in seahorses to the forefront, stimulated dialog and awareness, and led to the development of support materials (including identification guides, and guidelines for making Non Detriment Findings (NDFs) (See CITES and national legislation, P. 62).

OVERLEAF Project Seahorse team members at a CITES meeting in Asia.

CONSERVATION

SEAHORSES AS FLAGSHIPS

Seahorses are instantly recognizable and charismatic, and make wonderful flagship species for marine conservation. They live in some of the most highly impacted parts of the ocean—shallow waters near to land, coral reefs, seagrass beds, and mangrove forests. They can also draw attention to the mountains of by-catch that are brought up during destructive and wasteful trawl fishing, since seahorses often live in very similar areas to target species (most notably shrimp). Their appealing shape makes for ideal and eye-catching graphics, and their unusual biology is sure to capture people's imagination.

As Project Seahorse and SOS (Save Our Seahorses) Malaysia say: "Saving seahorses means saving the seas." This slogan goes two ways—to save seahorses requires us to save their habitats, and saving their habitats will save much more than seahorses alone. A focus on seahorses is also stimulating much-needed dialog in many areas of marine conservation, including the issues of wildlife trade, use of wildlife for magico-religious and traditional medicine purposes, undertaking species conservation assessments with limited data, by-catch, dredging, land-reclamation and coastal development, as well as illegal and unreported fishing and export. Seahorse researchers are also helping to lead the field in terms of potential conservation solutions, including establishing marine protected areas, using CITES as a management tool for marine fishes, the role of the public participating as citizen scientists in collecting meaningful conservation-related data, and the development of diving codes of conduct to avoid damage to marine ecosystems.

POPULATION DECLINES

Amanda Vincent first brought to public attention the fact that seahorse populations are declining. Reports from the central Philippines suggested that numbers were 50–70 percent lower in the mid-1990s than they had been 20 years earlier. Subsequent research by Amanda and her students and colleagues throughout Asia, east Africa, America, and Brazil confirmed that the problem was global in scope. Worryingly, the researchers also found new source countries (for example in west Africa and Latin America) entering the trade.

As populations decline, seahorses will have a harder time finding mates. Without intervention, this negative feedback loop continues, and replenishing the population becomes increasingly difficult. Further declines can lead to a population unable to bounce back to health, resulting in local extirpation, and, if unchecked and repeated in other such populations, this will eventually result in global extinction. As an interim conservation measure, it was recommended by the CITES Animals Committee in 2004 that seahorse trade be limited to individuals greater than a universal minimum size limit of 4 inches (10 cm) in height (from coronet to tail tip). The reasoning was that by 4 inches (10 cm), individuals of most traded species have already reached sexual maturity, and hopefully have had a chance to breed, thus avoiding some of the risk of population decline.

We do not know the conservation status of most seahorse species because they either have not been, or cannot be assessed, but the ones that we can evaluate are threatened. The International

IUCN REDLIST CATEGORIES

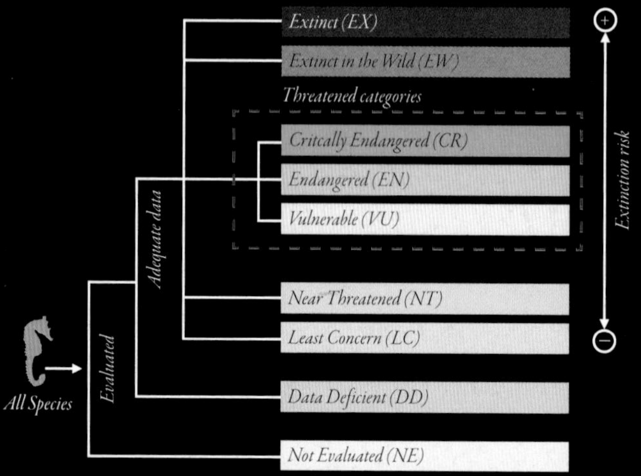

LEFT IUCN Redlist categories of conservation status for assessed species.

OVERLEAF Maximum size, and calculated size at first maturity for 32 seahorse species. Proposed Minimum Size Limit of 4 inches (10 cm) (marked in brown) for traded species falls above size at first maturity for the majority of species. (Redrawn from Foster & Vincent, 2005).

Union for the Conservation of Nature (IUCN) publishes a Red List of Threatened Species (www.iucnredlist.org) in which species are assessed according to a number of quantitative criteria related to population size and structure, geographic distribution, and changes in these over time. Based on these assessments, the conservation status for each extant species is determined: Least Concern (LC), Near Threatened (NT), Vulnerable (VU), Endangered (EN), or Critically Endangered (CR). Species for which insufficient data exist to make a reasonable assessment of their status are considered Data Deficient (DD). The 2015 IUCN Red List includes 27 seahorse species as DD, 11 as VU, one as EN, and one as LC. The rest (six) have not even been evaluated yet (Not Evaluated, NE).

The Red List has no legislative weight, however it does serve as a warning that species are in trouble. As such, it is theoretically used by governments and international bodies in prioritizing species for protection under their biodiversity-related laws. For seahorses, the large number of DD and NE species indicates that more research is urgently required in order to garner a fuller picture of their conservation status. Assessments are typically reviewed every five to ten years and are conducted by the relevant IUCN Species Specialist Group. The *Syngnathidae* fall under the remit of the Species Specialist Group for Seahorses, Pipefishes, and Sticklebacks, which is housed at Project Seahorse, and comprises 15 researchers, and aquarists with expertise in these fishes from nine different countries worldwide.

DESTRUCTIVE FISHING

Trawling is prevalent in countries all around the world, and is one of the most destructive and wasteful of fishing methods. Trawling usually focuses on a particular target species (for example, shrimp or prawns), but trawls are indiscriminate in what they take, and trawling is increasingly being used to capture any and all species, especially in south and Southeast Asia. Trawlers operate by dragging a net through the water, usually with a heavy beam holding it open at the front. Any benthic structure in the way of the trawl is flattened by the beam, and mobile creatures, regardless of species, are tossed up, and into the net. Approximately 67–73 percent of the catch in shrimp trawls,

for example, is non-target by-catch, commonly including seahorses and pipefishes. Since seahorses live in shallow seagrass and deeper water soft-bottom habitats where these fisheries generally operate, targeting shrimp is effectively similar to targeting seahorses.

Although few seahorses are caught by any one trawler (on average about one seahorse per trawler per day), the sheer scale of fishing worldwide adds up to more than 40 million seahorses per year (based on just 21 countries) that are inadvertently removed from their homes as by-catch.

Many fishers are aware that seahorses are valuable, and thus retain and sell them. However, even those seahorses that survive the catch and are thrown back in the sea are very likely to die as a result of stress or predation, or at the very least will have been displaced from their territory and their partner, and probably will not contribute to the next generation.

In addition to trawling, seahorses are also caught by means of a number of different fishing gears (including seines, traps, and gill-nets) and are impacted by other destructive fishing practices such as the use of dynamite to stun fish (which also inadvertently blows up and destroys the reef). However, these are less of an issue for seahorses than trawling, which accounts for nearly 80 percent of incidental captures and nearly 95 percent of the individuals that enter the trade. Seahorse conservation will only be truly effective as a result of substantial changes to the trawl industry. Can the charisma of seahorses help draw the world's attention to the terrible waste that is going on in these fisheries?

LEFT Trawling, where a weighted net is dragged along the seabed collecting nearly everything in its path, is one of the least-selective and most-destructive fishing methods.

LIVELIHOODS AND CONSERVATION

For many people, particularly in coastal areas of Southeast Asia, fishing is key to survival. Fish are not only their primary source of protein, but also their only source of income. With limited land or other resources, there are few alternative livelihood options for these people. Fishing is also frequently an occupation of last resort since the sea is basically "open access," meaning that even if they have nothing else, anyone can make use of marine resources.

Seahorses do not provide significant nutritional value for fishers and their families. However, their high market value means that many people do depend on them for a substantial fraction of their income. In the central Philippines in the 1990s, many fishers obtained 40 percent of their annual income (100 percent seasonally) from seahorses. It is unclear how recent trade regulation under CITES has impacted human livelihoods in the Philippines, and elsewhere that seahorses have historically been collected. However, in theory, improving sustainability of fisheries and trades should have a positive impact on the people that depend on them, enabling them to be able to continue fishing into the future.

When considering options for conservation, it is important to be aware of these human factors. Conservation measures are most likely to succeed if they take into account the needs and livelihoods of the people who depend on the resource requiring conservation, and if stakeholders themselves are involved in the development of the conservation measures (creating win-win situations for all involved). Conservation measures that necessitate changes to livelihoods should be accompanied by exploration of alternative, less damaging, livelihood options. In the central Philippines, Project Seahorse worked with villagers to develop craft-based income-generating projects, and other alternative livelihoods have included seaweed farming, eco-tourism, seahorse spotting as a dive attraction, and community-based aquaculture. Although these engaged people, and brought in some income, they did little to reduce fishing by the community as a whole, or alleviate pressure on the seahorses and further solutions are needed.

ABOVE A fisher in India checking his crab trap. Seahorses, drawn in by the shrimp bait, are often caught in the pots.

OVERLEAF Culturing seahorses is unlikely to make a huge impact on the traditional medicine trade, but it is important for the aquarium trade

AQUACULTURE

It has been suggested that captive breeding and culture of marine species (aquaculture) could help relieve the pressure caused by the direct exploitation of wild populations. That said, there are substantial difficulties that need to be overcome for commercial seahorse aquaculture ventures to be viable. As a result, the financial outlay and technical assistance required for setting up and maintaining a captive-breeding facility for seahorses is generally such that they cannot compete with the wide availability of cheaply obtained wild-caught (especially trawled) specimens, particularly for the TCM market.

The first commercial seahorse aquaculture project (to supply the TCM trade) dates back to the late 1950s in China. Since then, there have been numerous attempts to commercialize seahorse aquaculture, although these have met with limited success and most have failed within the first year. Breeding of seahorses for the aquarium market, where individual seahorses are more valuable than they are in TCM, is a more viable economic option, and many species are now available from entirely closed life cycles (i.e. seahorses are bred from young that were themselves born in aquaria). Seahorses from a company in Hawai'i retail today for as much as $450 each. An aquaculture facility in Sri Lanka easily demonstrates CITES requirements for captive-bred species by breeding a non-native species (*H. reidi* from the western Atlantic). For aquarium use, captive-bred individuals are preferable to wild-caught ones, having been raised on frozen food in a disease-free environment. The seahorses in zoos and aquaria worldwide mostly come from captive-breeding facilities and/or exchanges with other zoos and aquaria.

Aquaculture has also been heralded as an alternative livelihood option. For example in the Spermonde Archipelago, south Sulawesi, a demonstration project has recently been set up (with the help of Mars Symbioscience) to show how low-technology community-based aquaculture of *H. barbouri* can provide a steady income using seahorses. The success (or not) of this project remains to be seen.

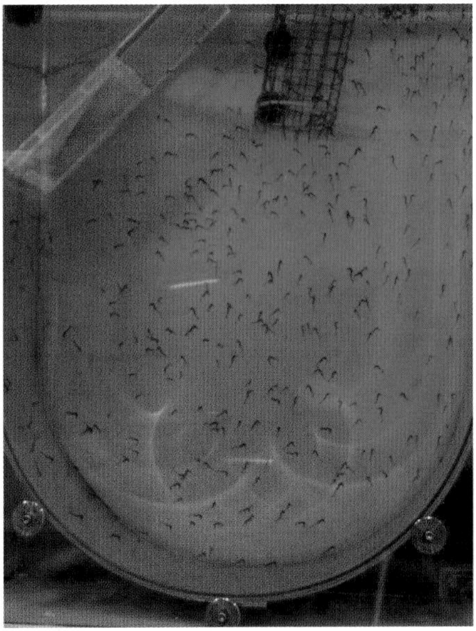

CITES AND NATIONAL LEGISLATION

Legislation can be a valuable conservation tool, formulated either on a country-by-country basis, or as the result of an international accord. For species that are threatened by international trade, the Convention on International Trade in Endangered Species (CITES) binds its 181 member countries ("Parties") to rules that regulate trade, and theoretically ensure that trade only takes place to the extent that is "not detrimental to wild populations". More than 35,000 species (mostly plants; only 103 fish) are on CITES and are listed in one of three Appendices. Appendix I is for species threatened with extinction and for which trade is essentially banned (for example, tigers), Appendix II includes species for which trade is controlled through export permits and quotas, and Appendix III is for species that are managed by particular countries but need the co-operation of other countries in order to achieve this management.

At the 12th Conference of the Parties to CITES (in 2002), all seahorses were listed on Appendix II. This was a landmark decision, as it was the first time that a marine species had been listed (with the exception of the Coelacanth in 1976). The listing was based on criteria similar to those used by the IUCN for their Red Lists (see Population declines, p. 56), and highlighted the need for trade controls to protect several over-exploited seahorse species. The entire genus was listed on CITES because of identification difficulties. Implementation of the listing, however, did not come into effect until May 2004.

Since 2004, Parties have been obliged to ensure that their seahorse exports are legal and sustainable, provide export permits, monitor any imports, and submit annual reports documenting their trade. What has happened in some countries, such as India, Malaysia and Indonesia (and until very recently, the Philippines), is that export of wild seahorses has been banned, thus avoiding the need to prove the sustainability of any export. That said, trade can still continue domestically, and also does occur internationally, but through unofficial and undocumented channels.

Concerns about the level of the trade in a particular species can generate a request for a "Review of Significant Trade" (RST). If this happens, exporting Parties must show evidence to back up the claim that their exports are "not detrimental to the persistence of wild populations". If, based on an evaluation called a "Non-Detriment Finding" (NDF), a particular Party is unable to defend such a claim, CITES will provide "Recommendations" to help them improve their trade, and has the authority to ban exports of that species from the Party in question as a last resort. Guidelines to help Parties carry out NDFs have been developed by Project Seahorse.

So far, two rounds of RSTs have been conducted. The first of these included three species: *H. kuda*, *H. kelloggi*, and *H. spinosissimus*; and the second included four species:

H. barbouri, H. histrix, H. trimaculatus, and *H. algiricus.* As a result of the findings from these studies, several Parties have received Recommendations that will hopefully help ensure that their trade becomes more sustainable in the future.

National or regional legislation that protects seahorses includes the EU Wildlife Trade Regulations (1996, 2004), the UK Wildlife and Countryside Protection Act (1981), the Slovenian Protection of Threatened Animals Act (1993), the Australian Environmental Protection and Biodiversity Conservation Act (1999), the New South Wales Fisheries Management Act (1994, 2004), and India's Wildlife Protection Act (2001).

The efficacy of these legislations, of course, depends heavily on the degree to which the fishers, traders, and fisheries and Customs officials are aware of the regulations, their willingness (and capacity) to undertake actions to comply with the quotas and to complete the paperwork, and the effectiveness of enforcement. In some places laws are deliberately not upheld, or they are undermined by corruption. In many others there are simply not enough financial or human resources (or the will) to bring perpetrators of wildlife crimes to justice.

Despite the limitations, the listing of seahorses on CITES has set a major precedent for marine taxa, and provides an interesting case study of how CITES can help with marine species conservation.

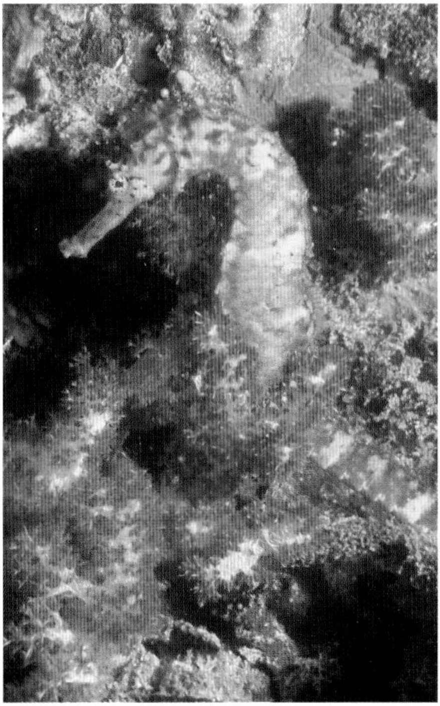

ABOVE TOP *H. comes*, the Tiger Tail Seahorse, has been the focus of a directed seahorse fishery in the central Philippines since at least the 1960s.

ABOVE Project Seahorse founder, Amanda Vincent, measuring seahorses by lantern.

SEAHORSE AMBASSADORS

At a more grass-roots level, dedicated seahorse supporters worldwide seek to understand the lives of seahorses, share their knowledge, and advocate for their conservation in many ways. These include divers, students, photographers, aquarists, writers, and conservation activists. Some that deserve particular mention include Australian photographer Rudie Kuiter, who has published several books and papers that amply demonstrate the incredible diversity of seahorses and their relatives. The British naturalist and author Neil Garrick-Maidment set up a seahorse-breeding program at the National Marine Aquarium in Plymouth, UK, and in the early 1990s, initiated the Great British Seahorse Survey, which has recorded thousands of seahorse sightings since it began in 1994. He also founded the Seahorse Trust (www.theseahorsetrust.org) in 1999, and the Seahorse Alliance (www.saveourseahorses.org) in 2011. Choo Chee Kuang, an energetic and dedicated researcher and advocate for seahorse conservation in Malaysia, set up SOS (Save Our Seahorses) Malaysia in 2004 (www.sosmalaysia.org), work that has been expertly continued by Adam Lim since Chee Kuang's untimely death in 2013. The late Denise Tackett was a passionate underwater photographer who brought *H. denise* to scientific attention and undertook extensive behavioral observations of pygmy seahorses. Paul Ferber spearheaded seahorse research and conservation work with Marine Conservation Cambodia (www.marineconservationcambodia.org). Louw Claassens has worked with the Knysna Basin Project (www.knysnabasinproject.co.za). Gaye Rosier started the Seahorse Project in 2014 to survey *H. guttulatus* with volunteers at Kenna Eco Diving (www.kennaecodiving.net). Rosana Silveira is involved with Projecto Hippocampus (www.projetohippocampus.org) in Brazil. Helen Scales wrote *Poseidon's Steed*, an extremely comprehensive popular book on seahorses, and Tami Weis writes many popular, and informative articles about seahorse husbandry and other topics on the web (www.fusedjaw.com).

Organisations involved with seahorse conservation include Zavora Marine Lab and All Out Africa, who are partnering

ABOVE LEFT Researchers from Project Seahorse checking seahorse populations in the marine protected area in Handumon, Central Philippines.

to monitor seahorses in Mozambique (www.zavoralab.com), and Peau-Bleue Association's EnQuete d'Hippocampe (Seahorse Quest) which started in 2005 (www.peaubleue.org). To this must be added many undergraduate, graduate, post-doctoral and professional researchers and aquarists who have been involved in elucidating the secrets of seahorses across the globe incuding Julia Baum, Elanor Bell, J.T. Boehm, Steve Casey, Miguel Correira, Janelle Curtis, Thelma Dias, Sarah Foster, Healy Hamilton, Dave Harasti, Suresh Job, Heather Koldewey, Aaron Lipton, Jackie Lockyear, Kristin Lunn, Keith Martin-Smith, Heather Masonjones, Aileen Maypa, Jana McPherson, Jessica Meeuwig, Marie-Annick Moreau, Sian Morgan, Arumugan Murugan, Marivic Pajaro, Nelson Perante, Alison Perry, Ierece Rosa, Peter Teske, Muthusamy Thangaraj, Amanda Vincent, Lucy Woodall, and all of Project Seahorse's current students, to name just a few. This list is in no way complete, and there are many other advocates who care about, learn about, and fight for seahorses and their ocean habitats, all of whom should be celebrated.

Undoubtedly one of the main conservation organizations focused on seahorses is Project Seahorse (www.projectseahorse.org). Project Seahorse was founded in 1996, as an international marine conservation and research organization. Amanda Vincent, who studied the behavioral ecology of seahorses for her PhD at Cambridge University in England, was the first to uncover the huge international trade in seahorses, particularly in Asia. What she knew of seahorse behavior and life history did not bode well for the sustainability of the trade, and she resolved to do something to help. Support from *National Geographic* in the early 1990s allowed her to travel extensively throughout Asia researching this problem, and this confirmed and worsened her fears.

A year later she had teamed up with Heather Koldewey and Helen Stanley at the Zoological Society of London (UK), and was back in the Philippines, with Marivic Pajaro from the Haribon Foundation in Manila, discussing options for a conservation project based on an island northwest of Bohol in the Central Visayas. Villagers in Handumon routinely collected seahorses at night by hand, to sell into the aquarium and traditional medicine trade. They had noticed that seahorses were becoming harder to find, and were as keen as the conservationists to ensure that their supply of seahorses did not run out.

Project Seahorse put much effort into developing participatory, community-based conservation solutions to try to mitigate the huge decline in seahorses that had been witnessed over the preceding two decades.

PROJECT SEAHORSE
advancing marine conservation

The fishers in Handumon were receptive to the idea of conservation action for seahorses and agreed to several measures, including building a "pregnant male cage." This was a wood and net enclosure anchored just offshore into which fishers placed captured pregnant seahorses. The males gave birth inside the cage and the young were able to escape through the net. This way, even though the father was sold, at least the young remained in the sea to contribute toward the next generation. Although this project worked technically, the logistics of maintaining the cage, and keeping track of whose seahorse was whose, led to conflicts and ultimately to its abandonment. In a separate initiative, the villagers designated a 120 acre (50-ha) area as being off-limits to fishing of all sorts—a marine protected area (MPA) or reserve. In the 19 years since the project began, the number of seahorses has increased within the MPA. Not only is this good for those seahorses within the MPA, but mating seahorses within the MPA also act as a source of recruits for the surrounding areas. The MPA has helped many other marine species increase their populations as well.

Project Seahorse has since grown to be a global organization, and has spawned more than 35 marine protected areas across the Visayas, as well as a local organization of fishers concerned with conservation of their local resources, from Handumon and a number of other communities on the Danajon Bank. It has worked closely with projects in Vietnam, Hong Kong SAR, and a score of other countries around the world. Research carried out by a succession of graduate students and others under Amanda's supervision has helped us understand so much more about seahorses; resource use, behavior, taxonomy, genetic connections, life histories, global-trade volumes, and routes, and contributed directly to conservation action plans.

Today, Project Seahorse is still based at the University of British Columbia in Vancouver, Canada, and the Zoological Society of London (UK). Guylian Chocolates from Belgium continues to be a major supporter. The project in the Philippines has matured into its own NGO, the Project Seahorse Foundation for Marine Conservation, and then merged with ZSL Philippines for ease of management. Thanks to Project Seahorse and many others, the plight of seahorses has entered the global consciousness. The dedicated people who work with Project Seahorse strive to realize the vision of "a world in which seahorses, and their relatives, are secure in well-managed ecosystems." They use seahorses as flagships for marine conservation, and approach the issue with a multi-pronged strategy, working with fishers, traders, students, and government officials to undertake and support research, assessments, legislation, education, and community based action and ecosystem protection.

iSEAHORSE

Mobilizing the efforts of seahorse enthusiasts around the world, Project Seahorse in partnership with iNaturalist, the University of British Columbia, the Zoological Society of London, Shedd Aquarium, and Guylian Chocolates, is now engaging the public as citizen scientists to help make thoughtful, appropriate, and effective conservation decisions regarding seahorses. They are doing this through iSeahorse (www.iseahorse.org), a smartphone app and online portal for citizen scientists that acts as a clearing-house for seahorse information, and also as a developing network of concerned citizens that are actively trying to protect our global oceans. iSeahorse provides users with access to regional identification guides, helpful protocols for undertaking surveys, and a place for users to contribute their own seahorse sightings and photographs and learn about how to set up "trends monitoring" in order to gather data on particular populations over time. The real goal, however, is to support contributors to take conservation action based on the data they collect, and thus turn citizen scientists into citizen conservationists. A toolkit to help users do just that will be part of iSeahorse soon, and the hope is that seahorse species are moved away from the Data Deficient and threatened categories of the IUCN Red List.

HOW YOU CAN HELP

Seahorses are clearly in trouble. Reversing the population declines will need a concerted effort of many people approaching the issue from many different angles. Here is a list of ways in which you can help seahorses and the habitats on which they depend.

- Avoid eating shrimp and make sustainable seafood choices
- Campaign for a reduction of trawling
- Raise awareness of seahorse conservation, and conservation of marine habitats in general
- Avoid purchasing wild-caught aquarium fish or curiosities
- Educate others
- Be careful when diving
- Contribute to marine conservation organizations
- Keep marine habitats healthy
- Get involved in citizen science projects such as iSeahorse
- Volunteer for marine conservation projects

iSeahorse.org

SEAHORSE SPECIES

INTRODUCTION

The species that are described in this book are those that are recognized in Lourie, Pollom, & Foster (in review) "Global Annotated Checklist of the Seahorses." Although it is acknowledged that there is some disagreement as to the exact definitions of some of these species, and that new discoveries continue to be made, this checklist is the most authoritative reference to date. It is based on extensive literature review, and examination of more than 2,000 seahorses in more than 28 different museums in seven different countries, including as many of the original type specimens as possible (those that were used by taxonomists to describe new species). It also incorporates genetic, photographic, and other evidence that has been brought to bear on this question in recent years.

Each species has its own page (or double page) with its scientific name and the author and date of its original description. Interesting information about each species is presented, including its most frequently used English common name, along with a fact box that covers some of the diagnostic characters that enable identification, a distribution map, depth, habitat, and conservation status according to the latest IUCN Red List assessment. Note that number of body rings are given as Number of Trunk Rings (TrR) + Number of Tail Rings (TaR). The most common value for rings and fin rays are given, followed by ranges in parentheses. The maps are a visual representations of the coastal areas that the species inhabit, however, they show a 200 m depth contour offshore, even though many species live at much shallower depths. The real range map for most species would result in such a narrow strip, particularly if trimmed to only include suitable habitat, that it would probably not even be visible on a map of this scale.

HOW THE SPECIES ARE ARRANGED

In the following pages, the species are arranged in groups (clades) according to their phylogenetic relationships (i.e. according to their genetic relatedness) (see Major clades, p. 46–47, and the consensus tree of relatedness) based primarily on work by Steve Casey, Peter Teske, and Healy Hamilton. Those species for which genetic data are not yet available, and/or those whose phylogenetic placement is unclear, are grouped together at the end.

PYGMY SEAHORSES

Pygmy seahorses differ from the larger seahorses, not only because of their size, less than 1 inch (2.5 cm) tall, but also in some diagnostic anatomical respects. For example, their skeletons are much reduced. Whereas the larger seahorses have solid ridges of bone surrounded by flat plates that articulate with, and slide over, one another, forming an almost impenetrable bony armor, all that remains of this armor in the pygmies are isolated crosses of bone embedded in the fleshy body. In addition, the young are carried not in a visible external pouch, but actually tucked up inside the male's body cavity. Unlike the Gastrophori pipefishes, which have a specialized external brooding area on their trunk, it appears that pygmy seahorses' brooding area is totally interior in placement, and its entrance may in fact still be posterior to the anus, suggesting that the trunk-brooding habit of pygmy seahorses and Gastrophori may have different origins. Further work is needed to elucidate the anatomical details.

Genetic data from pygmy seahorses thus far are limited, but initial results suggest that they are monophyletic (i.e. derive from a single common ancestor), form a sister group to the larger seahorses (i.e. are separate from them but most closely related to them), and branched off early from the main seahorse lineage.

The boom in SCUBA diving in recent years, and in particular the increasing interest among divers in macro-photography, is bringing more small animals, such as these tiny seahorses, to peoples' attention. The interest paid by divers does, however, have its downsides. Pygmy seahorses are sensitive to disturbance and to bright lights. They can show signs of stress, including unnatural movements, turning away, and retreating to their "core area" or behind branches of coral, as a result of direct or inadvertent harassment, such as physically touching the animal, moving them to get "the perfect shot," or using flashlights or flash photography near them. Richard Smith, who did his PhD on *H. denise*, has devised a set of recommendations and best practices , to ensure that divers get a chance to observe the pygmy seahorses that are often high on their "must see" list, while still being respectful of the animals and their habitats that they have come to visit.

H. bargibanti	*H. pontohi*
H. colemani	*H. satomiae*
H. denise	

TEMPERATE AUSTRALASIAN SPECIES

These two temperate seahorse species, found only in Australia and New Zealand, are each other's closest relatives (sister species), and branched off from the main seahorse lineage relatively early in its evolution. The fact that pygmy seahorses are also found in Australia (and these are evolutionarily even older) lends support to the idea that perhaps seahorses may have originally evolved in this part of the world (see Evolution, p. 44).

H. abdominalis	*H. breviceps*

H. HISTRIX **CLADE**

One of the largest sub-groups of seahorses is the "*H. histrix* clade." This group includes the famous *Hippocampus histrix*, whose name has been used indiscriminately for almost any spiny seahorse for decades. However, the name *H. histrix* is one that has been used/misused for almost any spiny seahorse for decades. Although the true H. histrix has a distinctive very long snout and sharp spines. There is debate as to the actual number of species in this clade (particularly in Australia), and exactly where the species boundaries lie. Further work is required to address this.

H. angustus	*H. jayakari*
H. barbouri	*H. subelongatus*
H. comes	*H. whitei*
H. histrix	

THREE-SPOT SEAHORSES

The next three species (*H. camelopardalis*, *H. planifrons*, and *H. trimaculatus*) can all potentially exhibit distinctive black marks on the back of their neck and trunk region, although not all individuals display them. It may be a gender-linked characteristic as males more commonly have the spots, while females often do not. The presence of spots may be an attempt to ward off predators, since they look like crab eyes when the seahorse is bent over at rest.

H. camelopardalis	*H. trimaculatus*
H. planifrons	

JAPANESE MINIATURES

Japan is home to a surprisingly wide diversity of seahorses, including a number of small species that are closely related to one another. Until the 1850s, Japan was essentially isolated from the rest of the world, and knowledge of its flora and fauna was similarly severely limited. This changed with the publication of *Fauna Japonica* (1833–1850), based on the collections belonging to the physician and naturalist Philipp Franz von Siebold, one of the few westerners allowed in Japan. This publication included descriptions of four seahorse species (*H. coronatus*, *H. brevirostris* (= *H. mohnikei*), *H. longirostris* (= *H. kuda*), *H. gracilissimus* (= *Acentronura gracilissima*), the latter is now classified as a pygmy pipehorse).

H. coronatus	*H. sindonis*
H. mohnikei	

SEMI-SPINY *H. KUDA* RELATIVES

Related to the global "*H. kuda* clade" (see below), is a group of species that show variable spine development and are found in the Indo-Pacific. Some are more or less smooth (especially the larger *H. kelloggi*), while others are distinctly spiny (especially small *H. spinosissimus*) and have been frequently misidentified as *H. histrix*. Some specimens have irregular spines, which may or may not represent separate species. As a result there is currently some debate about the species distinctions within this group, but for now we follow the "Global Annotated Checklist of the Seahorses" by Lourie, Pollom, & Foster (in review) and describe three species in this group.

H. borboniensis (tentative placement)	
H. kelloggi	*H. spinosissimus*

***HIPPOCAMPUS KUDA* CLADE**

The "*H. kuda* clade" has a global distribution. Species distinctions within the group are still debated, however most areas have genetically distinct local populations that may have arisen from rare founder events (for example after colonization by rafting individual(s) in floating mats of algae), and have very little communication with other such groups. As a whole, the species included in this group tend to be relatively smooth, with short to medium-length snouts. They are highly sought after for traditional medicine, mostly live in shallow, inshore habitats, and can survive in brackish water estuaries. (*H. borboniensis* may be part of this clade, see p. 109)

H. algiricus	*H. ingens*
H. capensis	*H. kuda*
H. fisheri	*H. reidi*
H. fuscus	

BASAL *KUDA*-OID SPECIES

Prior to the diversification of the previous two groups (*H. kuda*-clade and semi-spiny *H. kuda* relatives), there was apparently another evolutionary split, which separated the ancestor of these two groups from the ancestor of *Hippocampus guttulatus* (Long-snouted Seahorse). Since *H. guttulatus* is restricted to the Mediterranean and eastern Atlantic, while the others are predominantly in the Indo-Pacific, it is believed that this split took place during the closure of the (previously circumglobal) Tethys Seaway around 15 mya.

H. guttulatus

***HIPPOCAMPUS ERECTUS* CLADE**

A second group of seahorses that inhabits the Atlantic is called the "*H. erectus* clade," after one of its members, *H. erectus* (Lined Seahorse). This group includes three closely related species on both sides of the Atlantic, as well as a diminutive species that is restricted to the northern Caribbean (although the evidence for the inclusion of this species is less strong). Speciation events that took place within the *H. erectus* clade may be the result of changes in ocean circulation: i) strengthening of the Amazon outflow may have separated the ancestors of *H. erectus* (Caribbean) from the ancestors of *H. patagonicus*, about 5.27 mya, and ii) subsequent dispersal of *H. erectus* across the Atlantic (perhaps mediated by a stronger Gulf Stream) resulted in colonization of the eastern Atlantic and Mediterranean by the ancestor of *H. hippocampus*, about 3.35 mya.

The huge Amazonian freshwater outflow acts as a barrier for a number of marine species. *H. patagonicus* is found south of the outflow, and *H. erectus* is found to the north. That said, recently the true *H. erectus* has been found near Rio de Janeiro, suggesting it must have crossed the flow somehow at least once, and probably more than once, given the genetic diversity of the populations. Further research on this southern outpost of *H. erectus* is needed.

H. erectus	*H. patagonicus*
H. hippocampus	*H. zosterae*

SPECIES OF UNCERTAIN PLACEMENT

The following species lack genetic information, and their taxonomic position is not obvious based on their morphology. They are therefore all placed in this section sorted by relationships within this group that can be inferred based on morphology, for example *H. jugumus*, *H. pusillus*, and *H. tyro* probably constitute a group, as do *H. minotaur* and *H. paradoxus*.

H. debelius	*H. minotaur*
H. jugumus	*H. paradoxus*
H. pusillus	*H. zebra*
H. tyro	

Hippocampus bargibanti

WHITLEY, 1970

BARGIBANT'S SEAHORSE

ADULT HEIGHT
¾–1 inch (2–2.4 cm)

RINGS
12 + 31–32

PECTORAL FIN RAYS
10

DORSAL FIN RAYS
14

SPINES
irregular, bulbous tubercles

CORONET
rounded knob

COLOR/PATTERN
gray body with pink tubercles (or yellow, with orange tubercles)

NOTABLE ANATOMY
very fleshy head and body, no visible rings, very short snout

DEPTH
53–131 feet (16–40 m)

HABITAT
only found on gorgonian seafans of *Muricella* spp.

CONSERVATION STATUS
Data Deficient

The first specimens of Bargibant's Seahorse were discovered in 1969, clinging to a seafan that had been collected and brought into the Nouméa Aquarium, New Caledonia, by Georges Bargibant. Among the most highly camouflaged of all seahorse species, *H. bargibanti* has a striated gray body and bright pink (or yellow-orange) tubercles, which match the stems and closed polyps of its host seafans (*Muricella* spp.) precisely.

It has an extremely short snout, and distinctive bumps (tubercles) on the body. Males and females are almost identical, with only very minor differences in their genital region. Males have a vertical slit that opens into the body cavity inside which the young develop, while females have a raised ring that may act as, support or house, an ovipositor to deposit eggs into the male's belly.

The late Denise Tackett (to whom this book is dedicated) spent many hundreds of hours underwater observing and photographing *H. bargibanti* in Lembeh Strait, northern Sulawesi. Based on her research, and that of Richard Smith, we know that *H. bargibanti* is typically found in male–female pairs. Sometimes multiple pairs share a single host seafan (up to 14 pairs in one case), and sometimes single individuals (adults or juveniles) are found on their own fans. The population density of *H. bargibanti* is very low (~ 1 per 6,460 square feet/600 m^2). This in part reflects the low density of their host seafans (~1 per 3,230 square feet/300 m^2, only 20 percent of which are occupied by seahorses).

The life-span of *H. bargibanti* is unknown, but they grow very rapidly (observably so within two weeks as juveniles), and remain on their host fan for at least ten months. Brooding males and juveniles have been observed in March, June, August, and November. The first birth to be witnessed (as far as we know) was by Denise and myself in August 1999. The young were less than ⅛ inch (2 mm) long, and incredibly 34 of them were born over a period of about 15 minutes. They did not settle on the fan, but instead drifted out to sea.

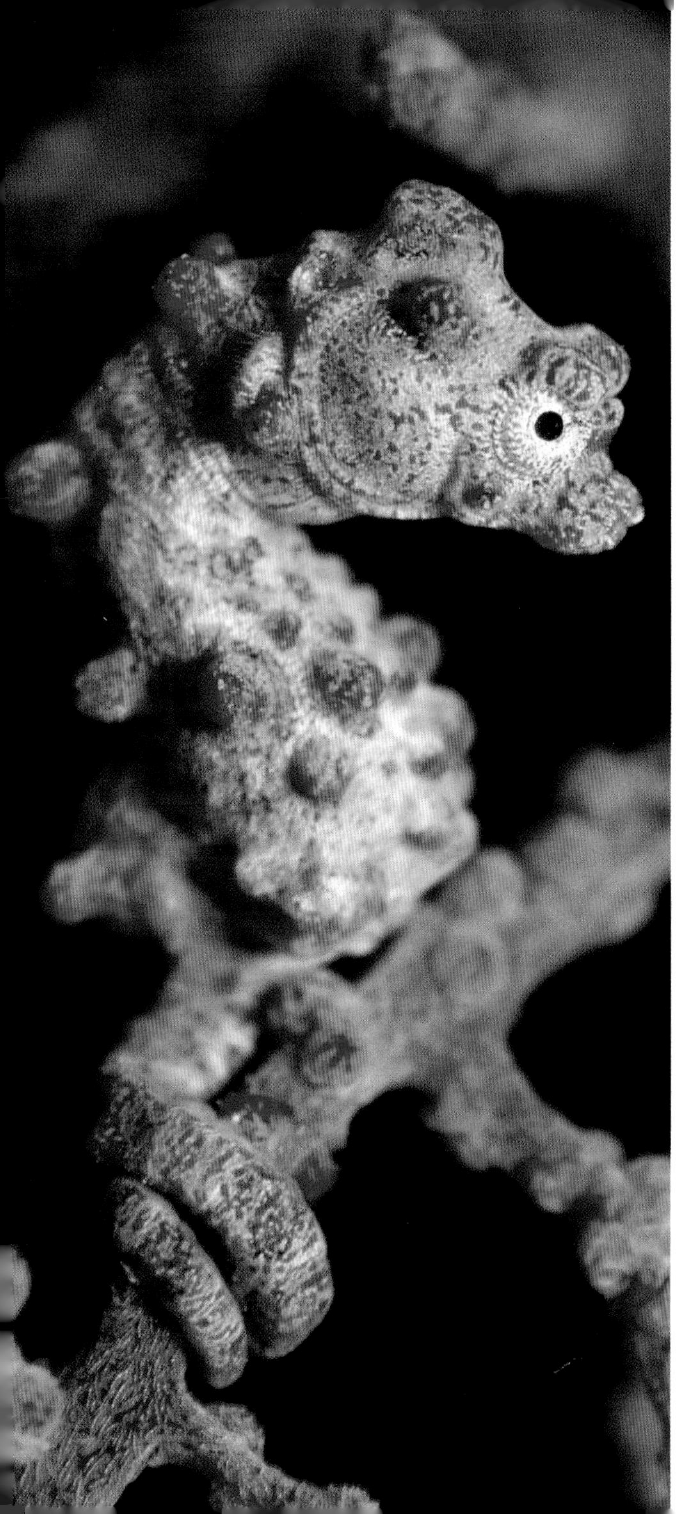

LEFT Bargibant's Seahorse is so perfectly camouflaged that it almost looks as though it is made out of the *Muricella* seafan on which it lives. It clings tightly to the branches and carefully uncurls and recurls its tail whenever it moves about the seafan.

ACTUAL SIZE

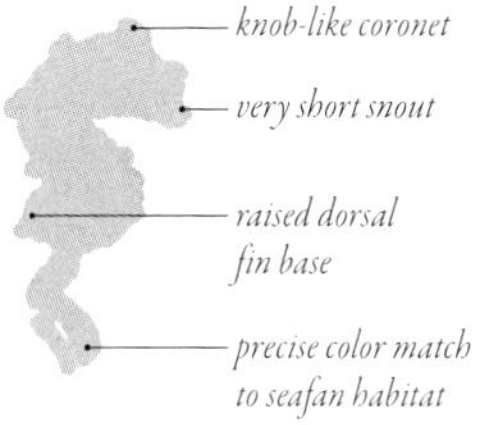

COLEMAN'S SEAHORSE

Coleman's Seahorse was named after underwater photographer Neville Coleman, who first discovered this species. It was found on Lord Howe Island, a remote and isolated volcanic crater rim situated off the east coast of Australia, beyond the Great Barrier Reef. Lord Howe Island harbors some very interesting marine life, much of which is endemic (found nowhere else on Earth), and this may be the case for *H. colemani.* That said, similar (but smaller) white pygmies have been found in Indonesia, the Philippines, and Papua New Guinea and are currently named *H. pontohi.* It may be that further research will show that these two are actually the same species.

Like the other pygmies, *H. colemani* has only a single gill opening at the back of its neck, and its young are carried within the trunk region. It has a very short snout, and distinctive fine red lines on its body. Unlike *H. bargibanti* and *H. denise*, it is not found in association with gorgonian seafans. The original type specimens were found in shallow water in an area of coarse sand with sparse seagrass (*Zostera* and *Halophila* spp.).

ADULT HEIGHT
½–1 inch (1.1–2.4 cm)

RINGS
12 + 28–30

PECTORAL FIN RAYS
10

DORSAL FIN RAYS
14

SPINES
relatively smooth, occasional small tubercles

CORONET
low, rounded

COLOR/PATTERN
pale golden yellow with fine red lines outlining patches of white

NOTABLE ANATOMY
single gill opening, posterior part of dorsal fin base strongly raised, fleshy head and body

DEPTH
17 feet (5 m)

HABITAT
coarse sand, *Halophila*, *Zostera* seagrass

CONSERVATION STATUS
Data Deficient

ACTUAL SIZE

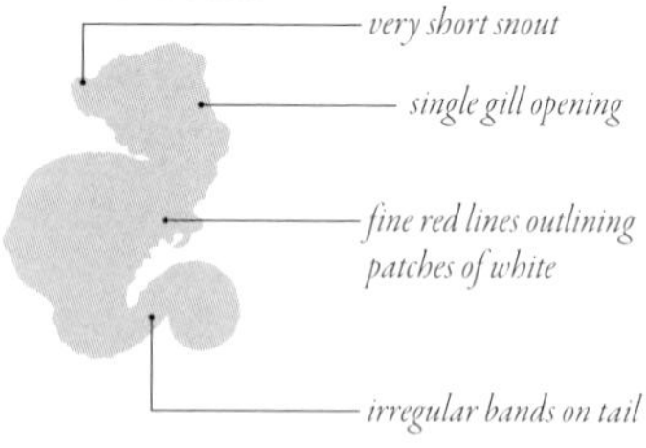

RIGHT Coleman's Seahorse is closely related to Pontoh's Seahorse and may prove to be the same species. However it is larger, more rounded, with a smooth head profile and was found initially in seagrass, whereas Pontoh's Seahorse is normally found on reefs in association with hydrozoans and bryozoans.

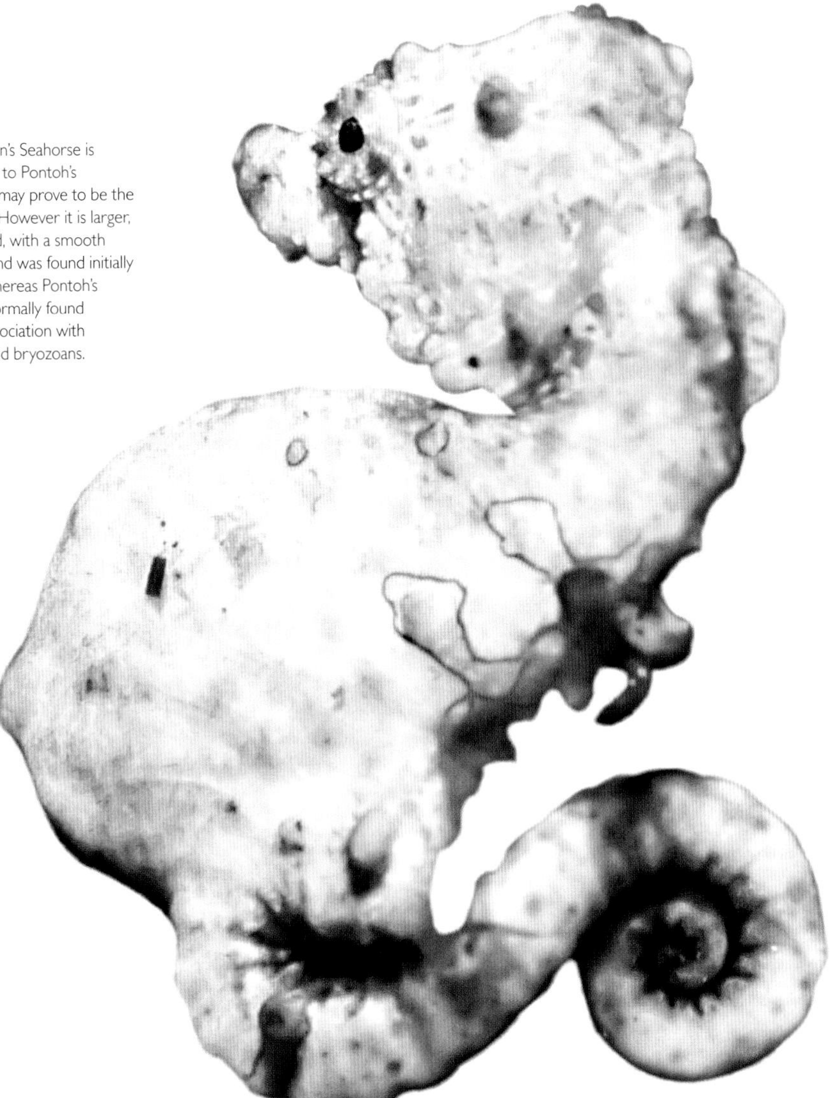

DENISE'S PYGMY SEAHORSE

ADULT HEIGHT
½–¾ inch (1.1–2.1 cm)

RINGS
12 + 28–29

PECTORAL FIN RAYS
10–11

DORSAL FIN RAYS
14

SPINES
smooth, or with occasional tubercles

CORONET
rounded hump

COLOR/PATTERN
pale orange with darker rings on tail, or reddish with white tubercles

NOTABLE ANATOMY
posterior part of dorsal fin base strongly raised, fleshy head and body

DEPTH
33–279 feet (10–85 m)

HABITAT
seafans especially *Annella* spp.

CONSERVATION STATUS
Data Deficient

Denise's Pygmy Seahorse was the second true pygmy seahorse species to be described, and is named after the late underwater photographer Denise Tackett who first brought it to my attention. It is more slender than *H. bargibanti*, and males and females differ significantly from one another (sexual dimorphism). The female is much longer and slimmer, and the male has a more rounded belly.

Like *H. bargibanti*, *H. denise* lives in obligate association with gorgonian seafans, especially *Annella reticulata*. However, it appears to be more flexible in its habitat choice and has been seen on various types of fan, including *Annella mollis*, and potentially species of *Muricella*, *Melithaea*, *Verucella*, *Acanthogorgia*, *Villigorgia*, and *Echinogorgia*. Seafan identification is very difficult for a non-specialist, especially in the field and without examining samples under a microscope, therefore some of these identifications are tentative.

H. denise forms pair bonds. Usually just one pair will inhabit a seafan, but one fan was host to six individuals. It is much more active than *H. bargibanti* and often detaches from the seafan, tucks its tail underneath its belly, and swims freely across the seafan. It is also more socially active. Individuals may interact closely with more than one partner, and in one case, a female *H. denise* was observed to maintain stable pair bonds with two males simultaneously. This is the first time polyandry has been reported in seahorses. The female impregnated the second male a mere three days after impregnating the first. Both males received only half a batch of eggs, however, the female presumably benefited by spreading her risk of reproduction between her partners.

Home range sizes vary from ~23–233 square inches (150–1,500 cm^2) on a single fan, and overlap significantly.

H. denise is primarily active during the day, with most social and reproductive activity occurring at dawn or dusk and feeding activity throughout the day. Pregnancy lasts 11–14 days, and a typical brood is 7–14 young (although may be as many as 30). Birth has been observed just before dawn (with 13 young expelled in a mere four minutes, and the male re-mating only 14 minutes after that).

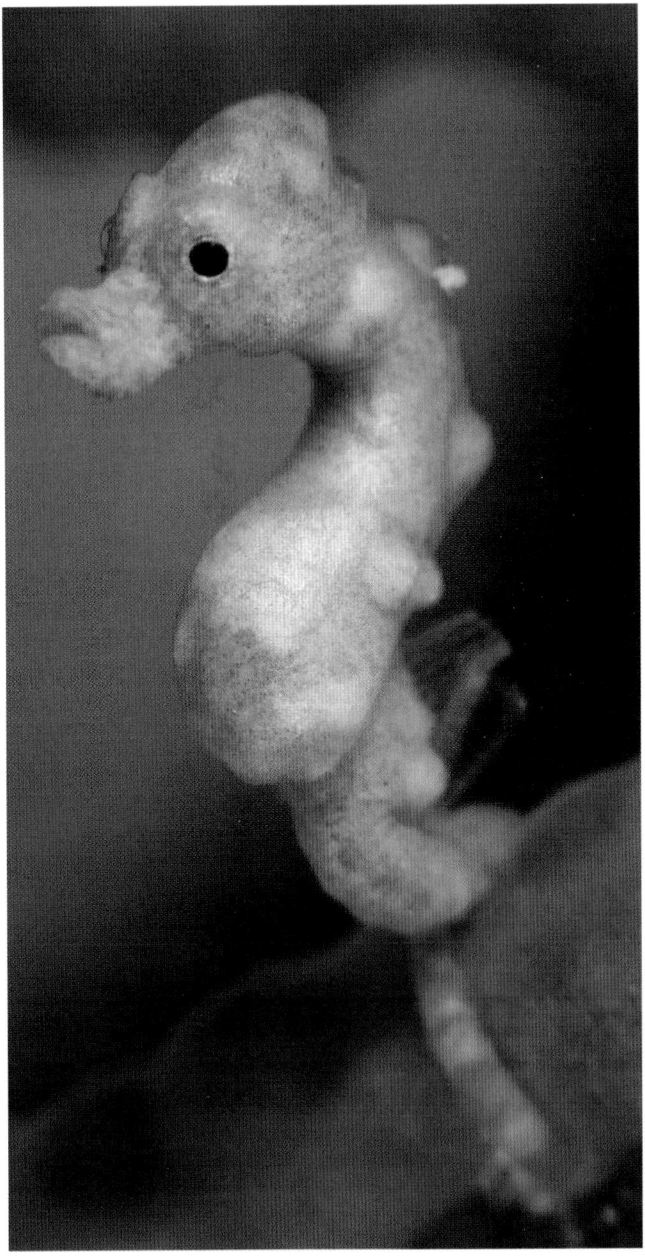

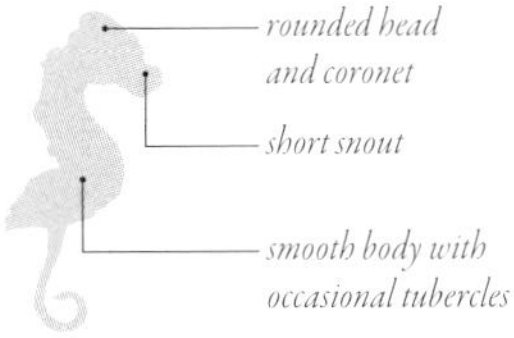

LEFT Denise's Pygmy Seahorse is smooth-bodied and slender. Its color and bumps match the seafans on which it lives. Being only ¾ inch (2 cm) tall it is easy to miss, although it is relatively active (at least compared to Bargibant's seahorse). This one was photographed in Halmahera, Indonesia.

PONTOH'S SEAHORSE

Pontoh's Seahorse is another recently described pygmy species. It is named after a dive guide in northern Sulawesi, Hence Pontoh, who has an impressive knack for spotting pygmy seahorses. *H. pontohi* is one of the most celebrated marine species of northern Sulawesi, appearing on numerous websites and attracting scores of divers.

H. pontohi was originally described with two other, very closely related, species. In fact, the only difference between two of them was their color—*H. pontohi* was white and *H. severnsi* was brown, often with a red patch over its dorsal surface. The hypothesis that they represented separate species was supported by the fact that both males and females could be either color, and that the two color varieties were never seen together. *H. kuda* also exists in two quite different colors: yellow and black. However, in this case, color seems to be sex-linked (females are more commonly yellow, and males more often black). Furthermore, mismatched *H. kuda* pairs are relatively common (the same is true for *H. comes*), and individuals can even change from yellow to black, or vice versa.

Since the publication of the original description, however, genetic data has indicated that *H. pontohi* and *H. severnsi* are in fact color morphs of the same species—their DNA sequences were identical. The International Rules of Zoological Nomenclature are clear: *H. pontohi* has precedence and *H. severnsi* is relegated to the file marked "synonyms."

ADULT HEIGHT
½ inch (1.3 cm)

RINGS
12 + 26–28

PECTORAL FIN RAYS
9–10

DORSAL FIN RAYS
14

SPINES
narrow, rounded, on head and dorsal ridge of certain trunk and tail rings

CORONET
slightly raised, angular

COLOR/PATTERN
white with fine red lines, and red bands across tail, or uniformly dark brown with red patch on dorsal trunk surface

NOTABLE ANATOMY
single gill opening

DEPTH
36–82 feet (11–25 m)

HABITAT
bryozoans, hydrozoans, coralline algae, gorgonian seafans, reef walls (or in fissures)

CONSERVATION STATUS
Data Deficient

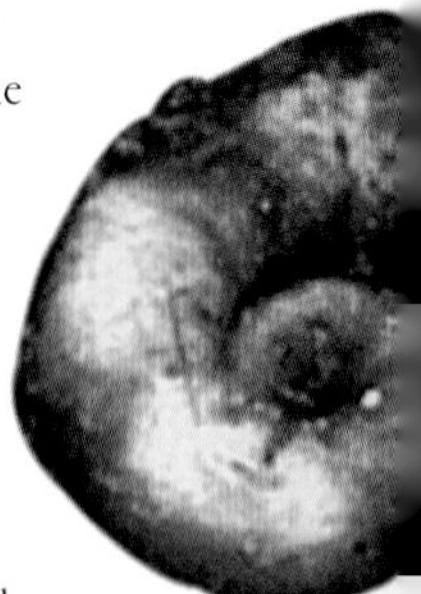

H. pontohi is found throughout Indonesia (both colors), and the brown form has also been found elsewhere in the Indo-West Pacific in medium depth water, among algae, bryozoans, seafans, and hydroids. Three pregnant males were collected in the months of June and July, though breeding could also take place at other times of year, and each had approximately 11 embryos. Further details of their life history are unknown, although, like *H. denise* they seem to be more active than *H. bargibanti.*

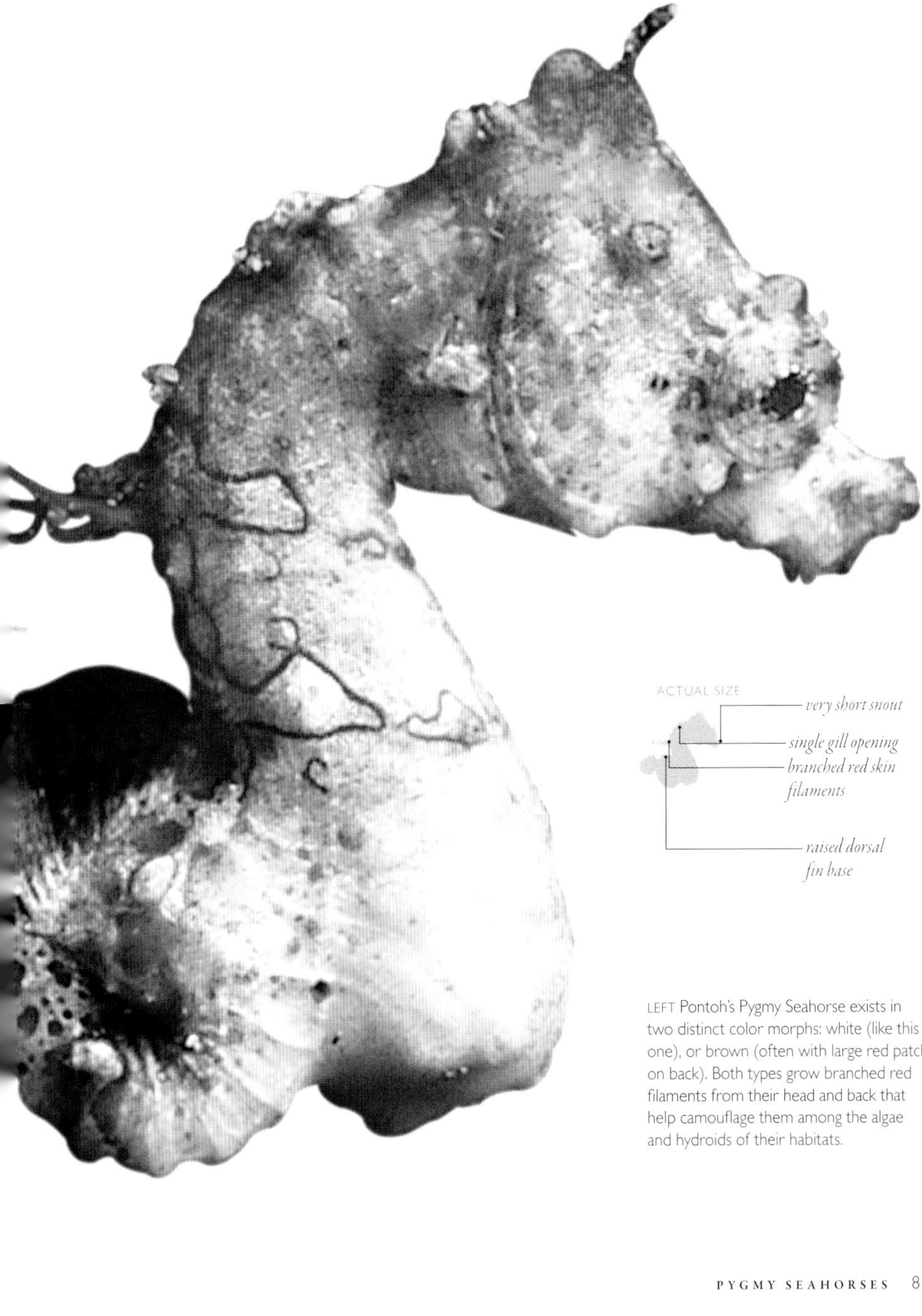

LEFT Pontoh's Pygmy Seahorse exists in two distinct color morphs: white (like this one), or brown (often with large red patch on back). Both types grow branched red filaments from their head and back that help camouflage them among the algae and hydroids of their habitats.

SATOMI'S SEAHORSE

Described in the same paper as *H. pontohi*, Satomi's Seahorse was first spotted by dive guide Satomi Onishi off Derawan in East Kalimantan, Indonesia. Only ½ inch (1.1 cm) tall, *H. satomiae* is the current contender for "World's Smallest Seahorse," and indeed one of the world's smallest vertebrates. The new discovery was celebrated as one of the "Top 10 New Species for 2008" by the International Institute for Species Exploration.

H. satomiae is distinguished from the other pygmies by the bony, square-tipped spines on its head, body, and dorsal tail ridge, which give it a distinctly rough appearance. It also has a black spot just in front of the eye and tiny white dots scattered all over its body.

At night, Satomi's Seahorses can be found congregating in small groups of three to five individuals, on seafans under overhangs. During the day, they are hard to spot, even in places where they are known to exist.

Birth has been observed on a number of occasions. The young are black, and approximately ⅛ inch (3 mm) long. The holotype specimen, collected in October, was brooding a total of eight young. Beyond the fact that the breeding season encompasses October, nothing more is known of *H. satomiae's* natural history.

ADULT HEIGHT
½ inch (1.1 cm)

RINGS
12 + 27–28

PECTORAL FIN RAYS
9

DORSAL FIN RAYS
13

SPINES
large, double spine above eye, prominent, rough spines on all body angles

CORONET
raised, H-shaped when viewed from above (laterally expanded anterior and posterior flanges)

COLOR/PATTERN
white, grayish, pale brown, a dark spot anterior to eye, blotchy red patches

NOTABLE ANATOMY
single gill opening, rough appearance due to spines

DEPTH
33–66 feet (10–20 m)

HABITAT
soft coral, gorgonians

CONSERVATION STATUS
Data Deficient

ACTUAL SIZE

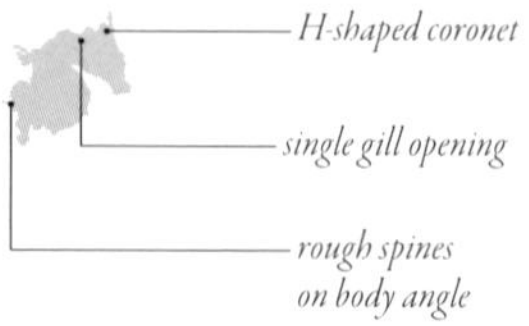

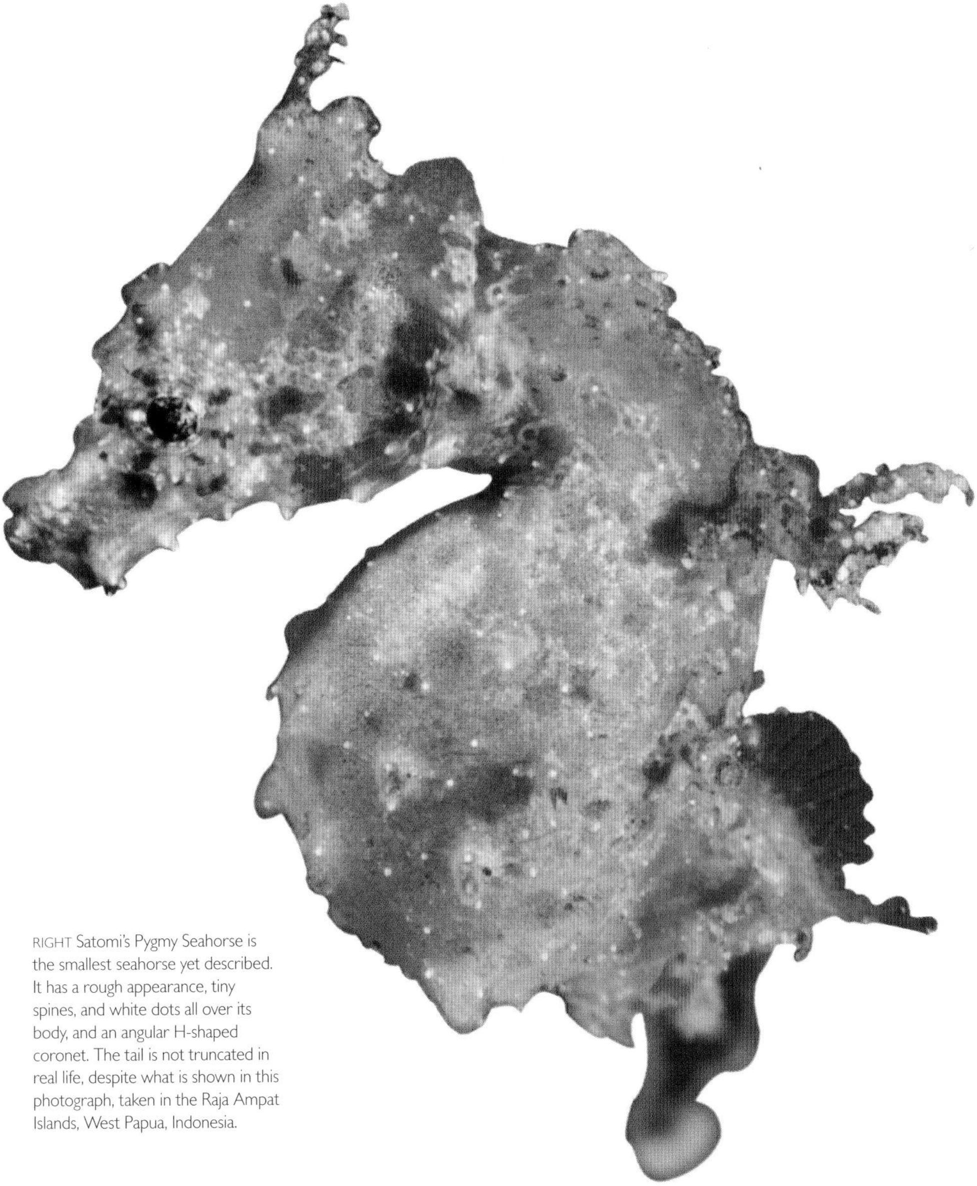

RIGHT Satomi's Pygmy Seahorse is the smallest seahorse yet described. It has a rough appearance, tiny spines, and white dots all over its body, and an angular H-shaped coronet. The tail is not truncated in real life, despite what is shown in this photograph, taken in the Raja Ampat Islands, West Papua, Indonesia.

BIG-BELLY SEAHORSE

ADULT HEIGHT
3¼–13¾ inches (8–35 cm)

RINGS
12–13 + 47 (45–48)

PECTORAL FIN RAYS
15–17

DORSAL FIN RAYS
27–28 (25–29)

SPINES
low, rounded bumps only

CORONET
low, triangular wedge

COLOR/PATTERN
pale white, yellow, or light brown, variably spotted especially on head and trunk (males more so than females)

NOTABLE ANATOMY
(very) prominent rounded eye spines, often with fronds on head, deep keel (males)

DEPTH
0–131 feet (0–40 m)

HABITAT
algae, seagrass, rocky reef, sponges, man-made structures

CONSERVATION STATUS
Data Deficient

The Big-belly Seahorse is one of the largest species, with full-grown individuals reaching heights of 13¾ inches (35 cm). Like the pygmies, it has 12 (or even 13) trunk rings. It is also the deepest-bodied seahorse, has the longest dorsal fin and the most tail rings. These higher counts may reflect a transition from a longer pipefish-like ancestor.

The Big-belly Seahorse is morphologically variable, particularly in terms of its snout length. This has led to questions about its taxonomic status. Rudie Kuiter recognizes as least two different species: *H. abdominalis* and *H. bleekeri*, based primarily on snout length differences. However, both long- and short-snouted specimens can be found in the same locations, even the same brood.

Occasionally specimens are found in South Australia, and the floating "Willyama seahorses" that were washed up *en masse* on South Australian beaches, in 2006, may be specimens of this species. It is also found on both North and South Islands in New Zealand.

H. abdominalis prefers low-exposure coastal habitats, such as bays, estuaries, and harbors. It uses macroalgae and man-made structures such as jetties, nets, and garbage as holdfasts. In general it is a shallow-water species, typically found in the intertidal zone to 131 feet (40 m), however, it has been recorded from a depth of 341 feet (104 m).

Genetically there is a distinction between Australian and New Zealand *H. abdominalis* populations (~1.4 percent cytochrome *b* sequence divergence). While this reflects lack of interbreeding, and is important from a management perspective, morphological variation is higher within individual populations than across populations, and the genetic divergence is considered too slight to warrant species status.

Breeding usually takes place in the Austral spring and summer, although the species can breed year-round. The main dispersal is believed to be a two to four week pelagic period after birth, during which young seahorses frequently attach to drifting algae. As pelagic juveniles, individuals maintain a more horizontal position, only becoming fully curled once they have settled as adults. Before this developmental change was understood, an entire new species (*H. graciliformis*) was described based on a juvenile *H. abdominalis*.

RIGHT One of the largest seahorse species, the Big-Belly Seahorse lives up to its name. Females usually have a very deep trunk region (like in this picture), and males can have enormous pouches. It has more trunk and tail rings than most species, and commonly has prominent filaments attached to its heads.

ACTUAL SIZE

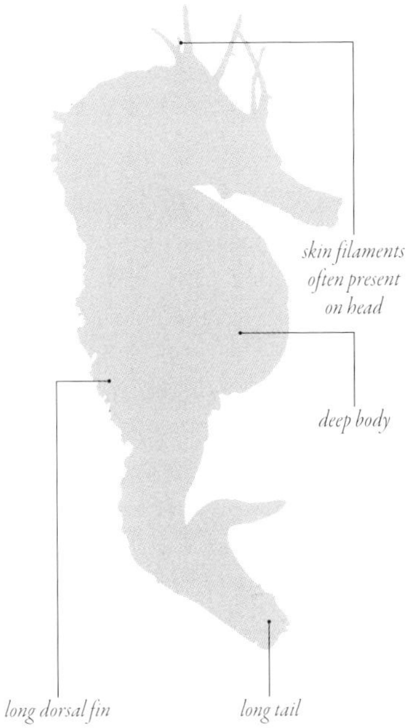

Hippocampus breviceps

PETERS, 1869

SHORT-HEADED SEAHORSE

The Short-headed Seahorse is most closely related to *H. abdominalis* (sister species), and it is possible that size-assortative mating may have given rise to the species distinctions i.e. *H. abdominalis* being the result of large seahorses mating with one another, and *H. breviceps* the result of small seahorses mating with one another. The Short-headed Seahorse is found in all southern Australian states, primarily in rocky reef/algal habitats, although there is some debate as to whether the populations in Western Australia represent a different species.

A study carried out in Port Phillip Bay, Victoria, in 2000, showed that *H. breviceps* lives in mixed-sex social groupings of two to ten individuals. This is unusual for seahorses. It is unknown whether or not the genetic mating structure is monogamous (i.e. one male and one female) as there is certainly opportunity for mate-switching. Socially, males were seen to display, and even interact, with up to three females. However, during the five-week study, each female only responded to the advances of a single male.

Individual seahorses were seen multiple times in the same spatial areas, suggesting some site fidelity. These spatial areas ranged in size from 10–129 square feet (1–12 m^2). Within the larger spatial areas, smaller core areas were used more extensively, and were the sites for most social interactions. As in other species, males tended to use smaller areas than females, perhaps because their movement is hindered by pregnancy.

ADULT HEIGHT
2–4 inches (5–10 cm)

RINGS
11 + 40 (39–43)

PECTORAL FIN RAYS
14–15 (13–15)

DORSAL FIN RAYS
20–21 (19–23)

SPINES
irregular, often prominent and rounded

CORONET
prominent, knob-like

COLOR/PATTERN
yellowish-brown to reddish-purple, with tiny dark-edged white dots

NOTABLE ANATOMY
short snout, commonly with stout filaments on head and trunk

DEPTH
0–49 feet (0–15 m); sometimes below in sponge beds

HABITAT
rocky reef, algae, seagrass beds

CONSERVATION STATUS
Data Deficient

Over two-thirds of the seahorses arrived and/or disappeared from the study site during the five-week study. This high turnover suggests that *H. breviceps* may be more mobile than many other species of seahorse, although disappearance from the study site could also reflect predation or death.

Density of *H. breviceps* in the algal beds was ~1 per 16–22 square feet (1.5–2 m^2). This is similar to densities found in other species such as *H. guttulatus*, *H. reidi*, and *H. capensis* in habitats where they are common. Sandy areas in between the algal beds did not harbor any seahorses.

LEFT The Short-headed Seahorse has a short snout, a column-like coronet and is covered with striking dark-edged white spots. It often has a 'mane' of thick filaments on its head and neck. It lives in algal beds and its distribution is restricted to southern Australia.

ACTUAL SIZE

Hippocampus angustus

GÜNTHER, 1870

NARROW-BELLIED SEAHORSE

The Narrow-bellied Seahorse was first described based on specimens that were collected from Shark Bay in Western Australia during the voyage of HMS *Herald* in 1858. There is some debate, however, as to whether the spiny seahorses in Shark Bay are members of the same species as those that occur further north in Western Australia, the Northern Territory, and northern Queensland. They could represent one end of a continuum of variation, or may be simply one of a number of independent species.

The northern Australian spiny seahorses tend to be variably mottled with dark brown markings on their bodies, and stripes across their snouts. The degree of development of the spines, the relative size of the head, the counts of tail and fin rays, and the length of the snout vary among individuals. Rudie Kuiter, in his revision of the Australian seahorses, suggests that a number of different species should be recognized—*H. grandiceps, H. hendriki,* and *H. multispinus* in addition to *H. angustus.* While acknowledging that morphological and genetic variation does exist in *H. angustus*, it is generally similar to that seen in other species of seahorses, and a clear pattern is not apparent. Therefore we follow the "Global Annotated Checklist of Seahorses" by Lourie, Pollom, & Foster (in review) and keep all these putative species under the name *H. angustus* for now.

ADULT HEIGHT
3–6¼ inches (8–16 cm)

RINGS
11 + 33–34 (32–35)

PECTORAL FIN RAYS
16–17 (15–19)

DORSAL FIN RAYS
18 (17–19)

SPINES
well-developed, blunt or sharp-tipped, usually low in neck region

CORONET
medium height with five well-developed spines

COLOR/PATTERN
covered with reticulating brown lines, striped snout, spines with dark band toward tip

NOTABLE ANATOMY
double sharp cheek spines, double spine below eye, sharp eye spine

DEPTH
10–207 feet (3–63 m)

HABITAT
unknown

CONSERVATION STATUS
Data Deficient

However, this conclusion may change in the future, and hopefully more genetic information will be published soon, which will shed additional light on the species question.

Many of the specimens of *H. angustus* were caught either by research trawls, or in the prawn trawling industry that operates off the north and northwest coasts of Australia. Apparently some of the seahorses came up from as deep as 164–197 feet (50–60 m). Nothing else is known about their ecology or behavior.

RIGHT The Narrow-bellied Seahorse is found in northwest and northern Australia. Most specimens have come up as by-catch from the prawn trawl industry. Variations among individuals have led to the proposal of different species names, but at present distinctions among the purported species are unclear.

ACTUAL SIZE

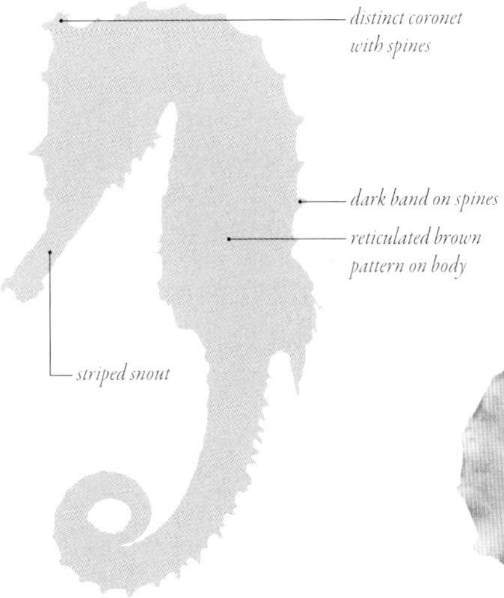

Hippocampus barbouri
JORDAN & RICHARDSON, 1908

ZEBRA-SNOUT SEAHORSE

The Zebra-snout Seahorse is a popular aquarium fish and marine curiosity, which has been commonly misidentified as *H. histrix* because of its spiny appearance. In the wild it is found in Malaysia (Sabah), the Philippines, and Indonesia, and it is closely related to the spiny striped-snouted seahorses from northern Australia, currently identified as *H. angustus*.

Historically, the main trade sources for this species were the Philippines and Indonesia. However, since the listing of seahorses on CITES in 2004, and the Philippines' subsequent domestic ban (now lifted) on seahorse exploitation, trade from there has diminished. Indonesia has also recently instigated a ban on export of wild-caught specimens. Fishers in Tanakeke, an island in the southern Spermonde Archipelago, off the southwest arm of Sulawesi, used to catch them with wooden nets pushed by hand through the seagrass in water just a few feet deep. They were sold to dealers in Makassar, and then shipped to Bali for international export.

In Pulau Badi, another island in the Spermonde Archipelago, a demonstration land-based aquaculture initiative provides a source for the aquarium trade, as well as a livelihood for local fishers. The Spermonde project has been underway since 2009, and the first captive-bred seahorses were exported in 2011 (2,800 CITES-certified seahorses from Pulau Badi were sold by January 2014).

ADULT HEIGHT
3¼–6 inches (8–15 cm)

RINGS
11 + 34–35 (33–36)

PECTORAL FIN RAYS
17–18 (15–20)

DORSAL FIN RAYS
19 (16–22)

SPINES
well-developed, sharp eye spine, first dorsal trunk spines longer and curved backward, tail spines alternately short and long

CORONET
medium height with five well-developed spines

COLOR/PATTERN
white to pale yellow, covered with reddish-brown dots and lines on body, snout usually striped

NOTABLE ANATOMY
sharp double cheek spines, double spine below eye

DEPTH
< 10 feet (3 m)

HABITAT
seagrass beds

CONSERVATION STATUS
Vulnerable

H. barbouri has been the focus of a number of research projects, investigating its feeding, breeding, hormones, and phylogeography (geographic distribution of genetic types). This research has shown that the species is active during the daytime, that gestation is mediated by some of the same hormones as in mammalian species, and that populations are generally genetically distinct from one another, suggesting limited dispersal capabilities.

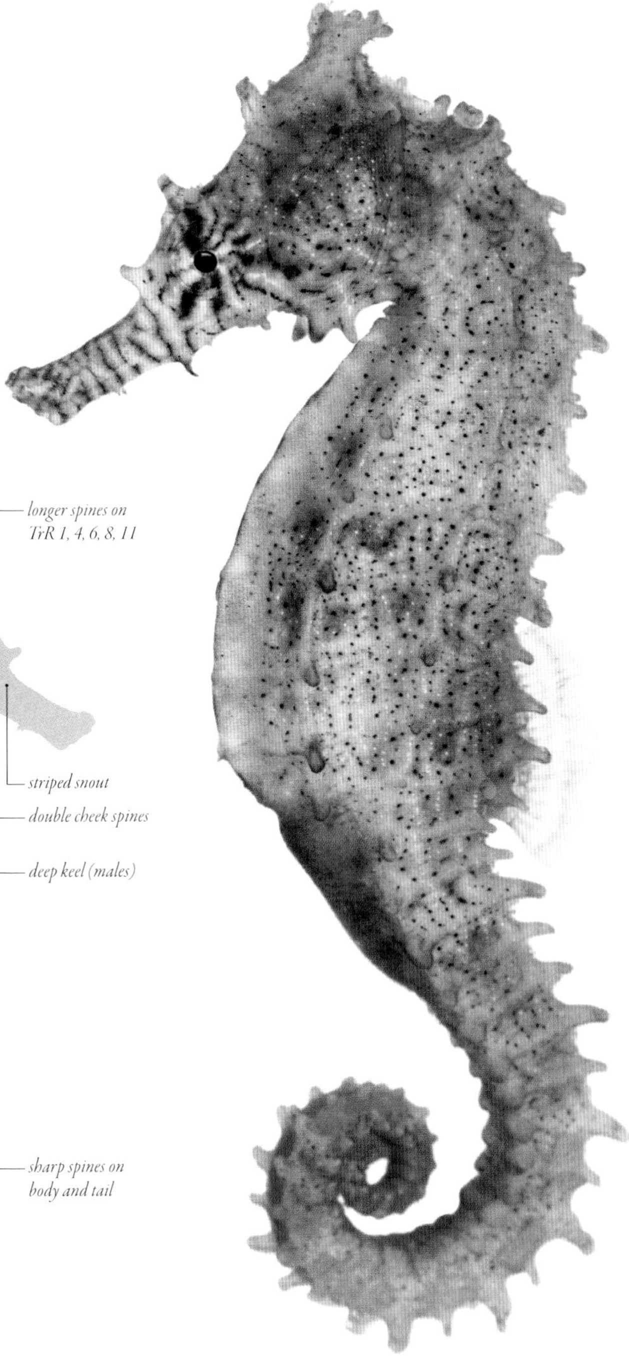

RIGHT The Zebra-snout Seahorse has a striped snout (as its name implies), a spiny exterior, and spotted brown markings all over its body and tail. The pattern of longer spines on its body (on trunk rings 1, 4, 6, 8, and 11) are diagnostic of the species and its closest relatives.

ACTUAL SIZE

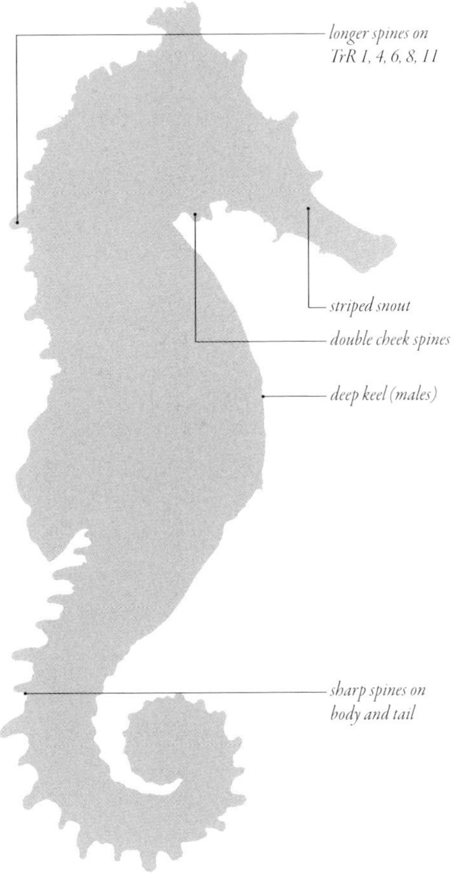

TIGER-TAIL SEAHORSE

The Tiger-tail Seahorse, so named because of its distinctive striped tail, was one of the first tropical seahorse species to be studied in detail in the wild. It had been the target of a fishery in Handumon (and elsewhere) in the central Philippines since the 1960s, and although collecting and trading seahorses in the Philippines was banned following the listing of seahorses on CITES, the trade continued as a result of poor enforcement and lack of support. Apparently the ban is no longer in place and so (sustainable) trade is now legal again.

The Tiger-tail Seahorse is found throughout Southeast Asia, particularly in the central Philippines, on coral reefs and in seaweed beds. It appears to be nocturnal. This is unusual for a seahorse, as most species seem to be more active in the day, with social activities, courtship, and mating occurring at dawn, and feeding taking place through most of the daylight hours. As adults, Tiger-tail Seahorses live in shallow coral reefs and spend most of the day hidden under the coral. At dusk they emerge and hang onto the coral branches (usually the same one every night) to engage in feeding and social activities. Typically two seahorses (presumably a mated pair) will share the same territory, and the same individuals have been seen in the same place for nearly two years, indicating pair-bonding and site fidelity.

Breeding in *H. comes* takes place year-round, with a peak in September–November, and a corresponding peak in recruitment (i.e. when young are old enough to settle in adult habitats) in February–March. Length of gestation varies during the year from an average of 13 days in March–August, to 20 days in the colder months of September to February. Average brood size is 400 young, and males can release their broods over multiple days. Juveniles are about 1 inch (2.5 cm) at settlement, and grow to reproductive maturity (about $4\frac{1}{4}$ inches/11 cm) within a year. Their longevity is estimated to be at least three years.

ADULT HEIGHT
$4\frac{1}{4}$–$7\frac{1}{2}$ inches (11–19 cm)

RINGS
11 + 35–36 (34–37)

PECTORAL FIN RAYS
17 (16–19)

DORSAL FIN RAYS
18 (17–19)

SPINES
knob-like, blunt (sharper in juveniles), often with dark band near tip

CORONET
low but distinct, with five rounded knobs or spines

COLOR/PATTERN
yellow to black with striped tail, mottled or blotched pattern on trunk

NOTABLE ANATOMY
striped tail, double cheek spines, low but distinct coronet

DEPTH
usually < 33 feet (10 m)

HABITAT
coral reefs (adults), *Sargassum* beds (adults and juveniles)

CONSERVATION STATUS
Vulnerable

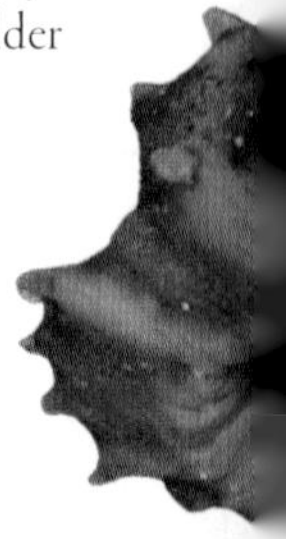

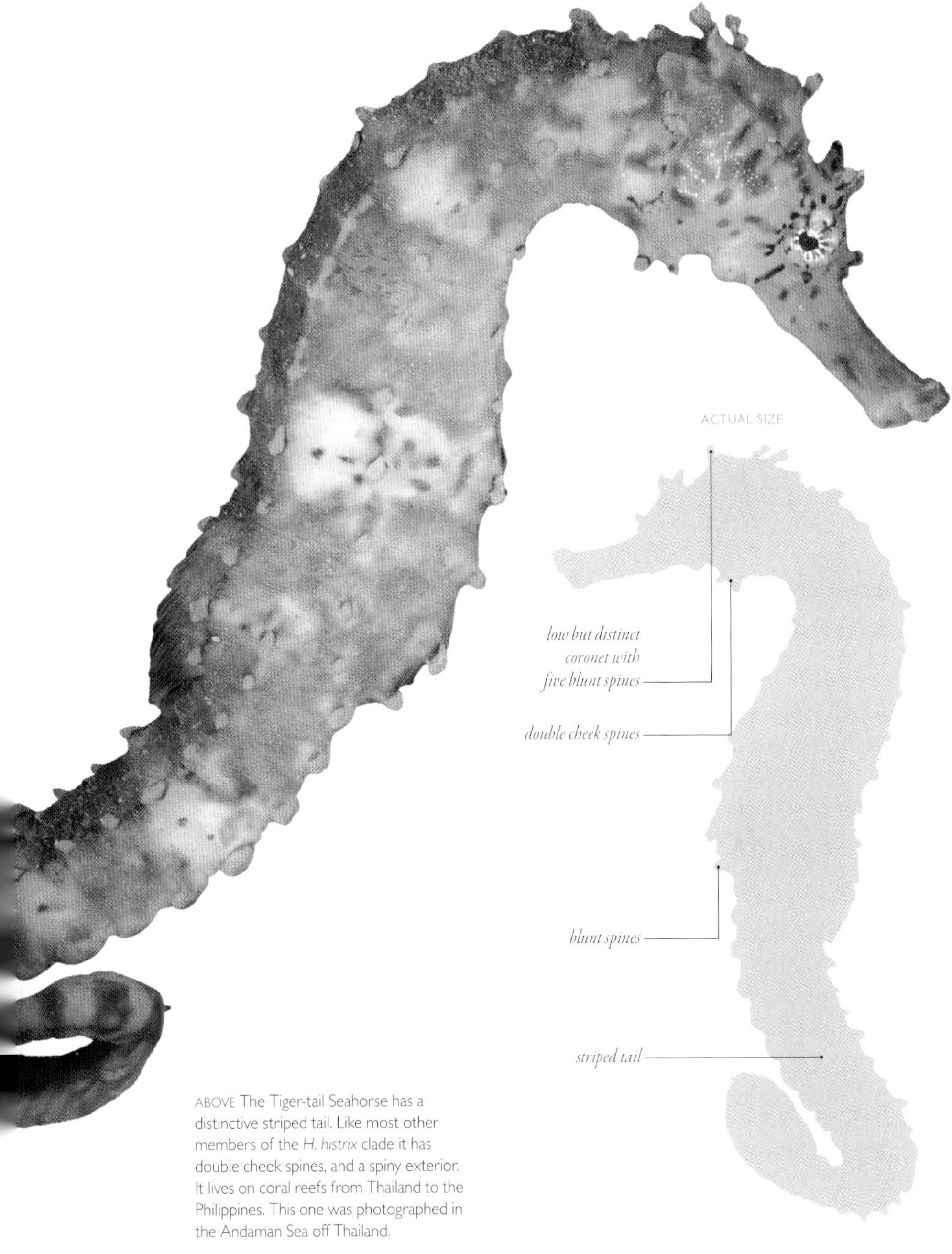

ABOVE The Tiger-tail Seahorse has a distinctive striped tail. Like most other members of the *H. histrix* clade it has double cheek spines, and a spiny exterior. It lives on coral reefs from Thailand to the Philippines. This one was photographed in the Andaman Sea off Thailand.

Hippocampus histrix

KAUP, 1856

THORNY SEAHORSE

The Thorny Seahorse is a highly distinctive species, with a very long snout, single cheek spines, a well-defined coronet with long spines, and long, dark-tipped spines on the trunk and the tail. Despite being so distinctive, its scientific name, *H. histrix*, has been misapplied widely, and used fairly indiscriminately for any spiny seahorses from the Indo-Pacific, regardless of their true species identity. This has caused much confusion in the literature, in trade records, in conservation planning, and in communication in general. Some of the species that have commonly been misidentified as *H. histrix* include *H. barbouri* and *H. spinosissimus.*

The geographic extent of the Thorny Seahorse stretches from east Africa through to Japan, New Caledonia, and Hawa'ii. Specimens from different parts of the range look remarkably similar, although having such a wide distribution raises the question as to whether this is in fact a single species. Initial genetic studies do show some deep divisions, suggesting that the current taxonomy may need some refinement, for example 6 percent divergence in cytochrome *b* sequence between east Africa and Japan. This is a much greater difference than is typically seen within a single species. More research is needed to shed light on this question.

Since the "true *H. histrix*" (as opposed to mis-identified *H. barbouri* and *H. spinosissimus*) is a relatively deep-water species (rarely found at depths of less than 33 feet/10 m), little is known about its biology and life history.

ADULT HEIGHT
3¼–6¾ inches (8–17 cm)

RINGS
11 + 35 (34–37)

PECTORAL FIN RAYS
18 (17–20)

DORSAL FIN RAYS
17 (15–18)

SPINES
extremely long and sharp, often black-tipped

CORONET
medium height, four to five long spines

COLOR/PATTERN
pale pink, yellow, green, may have mottled saddles across back, spines often dark-tipped

NOTABLE ANATOMY
very long snout (>½ head length), dorsal fin base very short, more pectoral than dorsal fin rays

DEPTH
usually > 66 feet (20 m)

HABITAT
sea pens, soft corals, sponges

CONSERVATION STATUS
Vulnerable

What is known is that it is commonly found in association with sponges, sea pens, or soft corals, where its pale pastel colors enable it to blend in masterfully. It is not a successful aquarium species, nor is it favored in the medicine market because of its spines. However, the move toward patented medicines in pill form, where the size and form of the powdered seahorse is not apparent to the purchaser, mean that even this seahorse is not immune to exploitation.

References to *H. histrix* in the literature, and elsewhere, should be treated with caution, unless accompanied by a photograph that clearly identifies it as the Thorny Seahorse.

RIGHT The very long snout immediately identifies this Thorny Seahorse. It also characteristically has very long, dark-tipped spines, and is frequently seen in habitats with soft corals or sponges, with which its (usually) pale pastel colors blend nicely. This photo was taken in Sumbawa, Indonesia.

ACTUAL SIZE

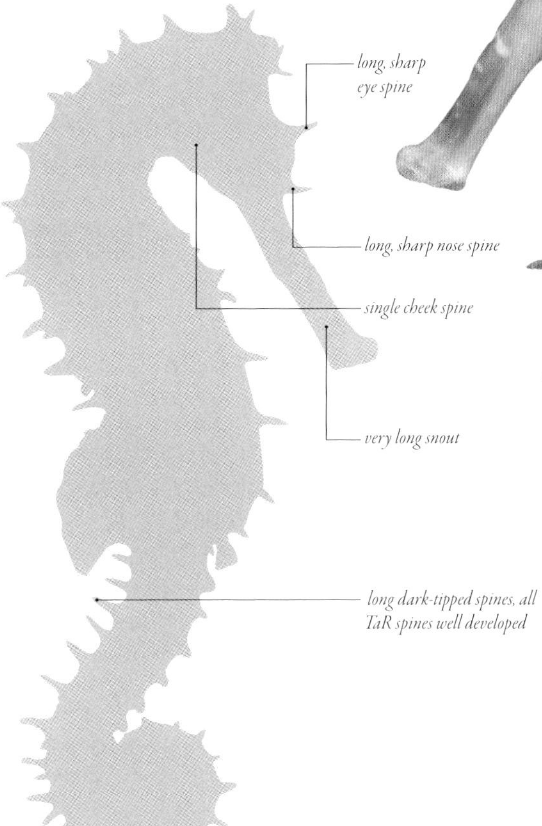

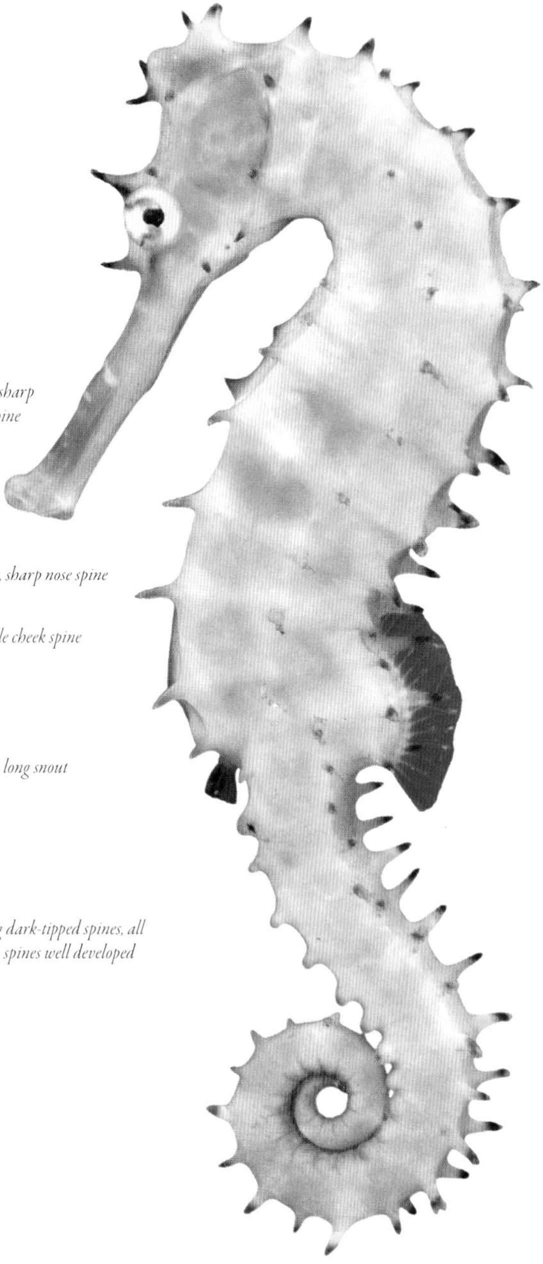

JAYAKAR'S SEAHORSE

Jayakar's Seahorse is closely related to the Thorny Seahorse (*H. histrix*), and takes its place in the Red Sea and the Arabian Gulf. These two species are probably each other's closest relatives, although no genetic data are presently available for *H. jayakari* with which to test this assumption. *H. jayakari* differs from *H. histrix* in tail ring and dorsal fin-ray counts, the form of the cheek spines, and the pattern of development of tail ring spines. It is also sometimes found with thick skin fronds attached to its head and neck (not seen in *H. histrix*), although this characteristic is not generally a good taxonomic character because skin fronds can be gained or lost depending on the seahorse's habitat.

Jayakar's Seahorse is most commonly seen in very shallow water, often in association with seagrass, unlike the Thorny Seahorse, which is a deeper-water species that prefers soft corals and sponges. The colors and spots on *H. jayakari's* body resemble algal growth on seagrass blades.

ADULT HEIGHT
4¼–5½ inches (11–14 cm)

RINGS
11 + 38–39

PECTORAL FIN RAYS
17–18

DORSAL FIN RAYS
18–19

SPINES
very long and sharp, alternate tail ring spines very low or lacking

CORONET
low-medium with four long, sharp spines

COLOR/PATTERN
pale cream or beige, large white spots on head and trunk, spines dark-tipped, or with dark band near tip, dark mid-ventral line

NOTABLE ANATOMY
alternate tail ring spines often poorly developed

DEPTH
7–10 feet (2–3 m)

HABITAT
Halophila, seagrass beds

CONSERVATION STATUS
Data Deficient

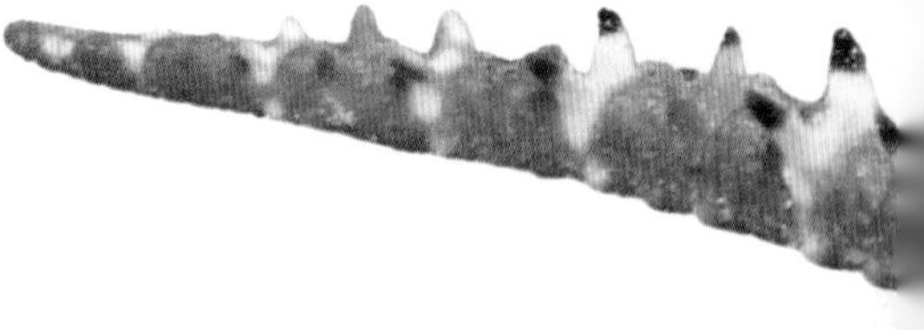

ACTUAL SIZE

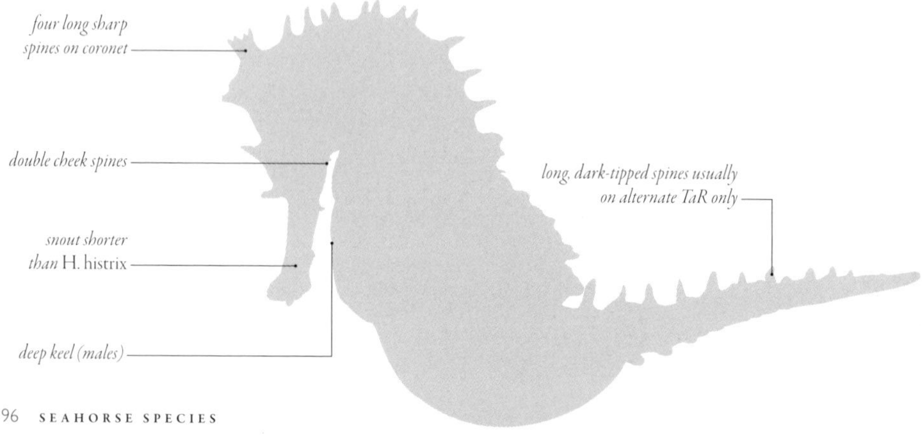

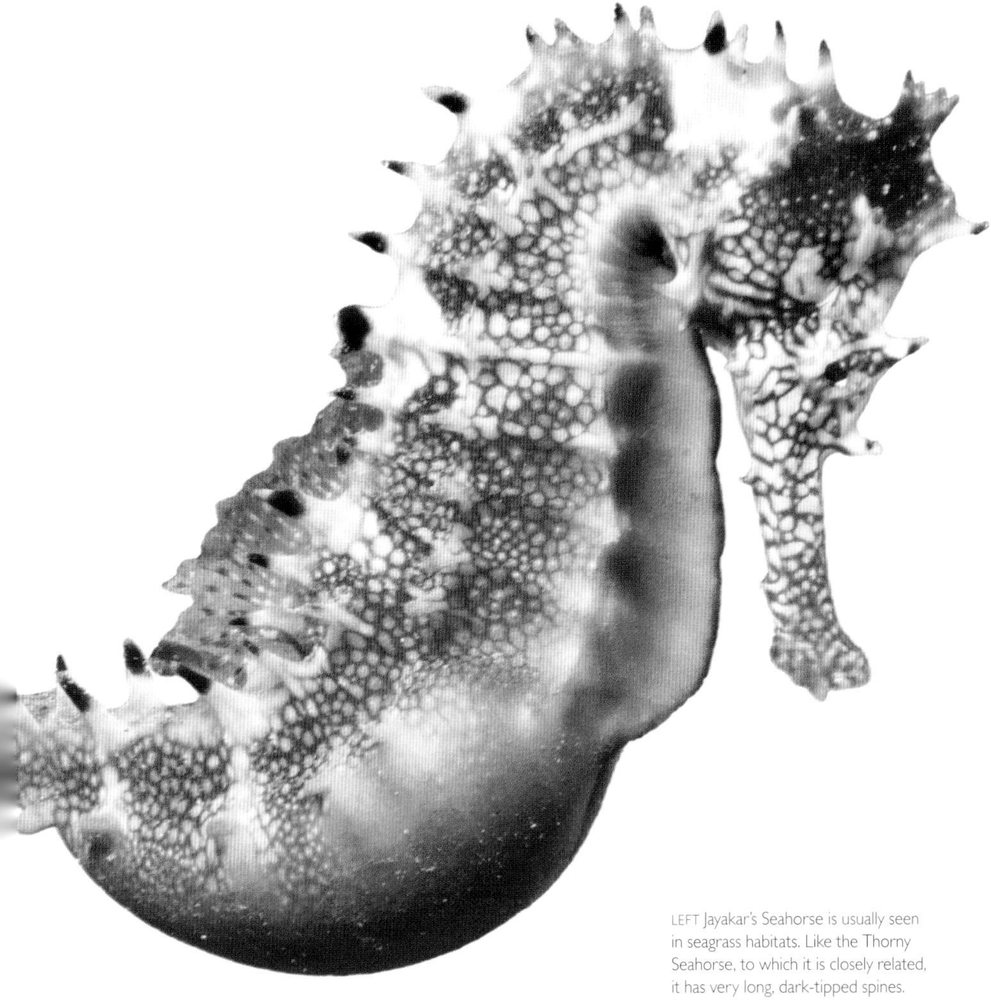

LEFT Jayakar's Seahorse is usually seen in seagrass habitats. Like the Thorny Seahorse, to which it is closely related, it has very long, dark-tipped spines. However, its snout is much shorter, and not all spines are developed, especially on the tail. It is often sandy-colored with white spots that camouflage it well.

Hippocampus subelongatus

(CASTELNAU, 1873)

WEST AUSTRALIAN SEAHORSE

The West Australian Seahorse is the most southerly species of the "*H. histrix* clade," and is not spiny at all. In fact most seahorse species that are found in temperate waters are not particularly spiny. It is possible that spines have developed only in the more diverse warmer regions as a protection against predation. The West Australian Seahorse has a distinctive, very high coronet with a rounded knob at the tip. Apart from its lack of spines, and very high coronet, the West Australian Seahorse is very similar to its more northerly cousin, the Narrow-bellied Seahorse. Further research is required to understand the genetic relationships between these species, and among all of the northern Australian spiny seahorses.

The West Australian Seahorse has a limited distribution, being found only in the southwest corner of Australia. Individuals typically live in muddy habitats, attached to sponges, sea squirts, or man-made objects (like ropes or jetty piles), or simply lying in depressions in the sand. They are tolerant of brackish water and can even be found quite high up the Swan River in Perth.

Researchers in Australia investigated the mating system of *H. subelongatus*. Using genetic information, they showed that, contrary to popular belief about seahorses and their faithful pair bonds, this particular seahorse species engages in what is called "sequential polyandry." This means that between breeding cycles mate-switching sometimes takes place (even when the original partners are available), and a female may deposit her eggs in the pouch of a different male. They are, however, faithful within a single breeding cycle as they cannot split a brood, and all the young from a single clutch come from a single female.

ACTUAL SIZE

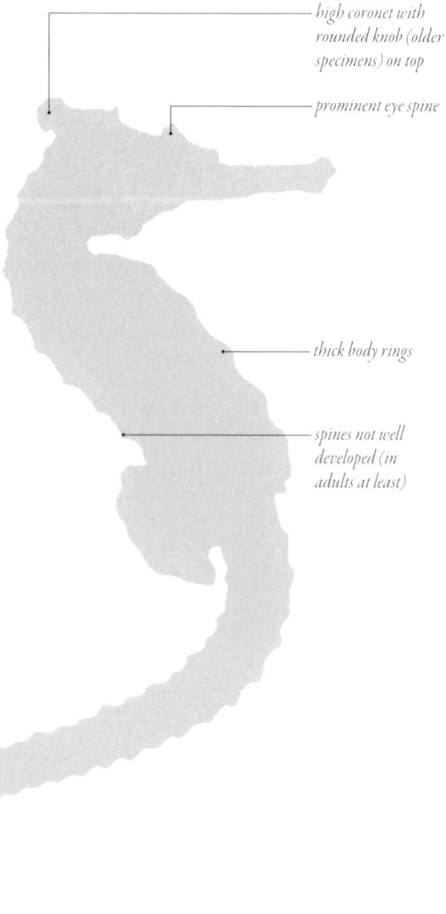

ADULT HEIGHT
5–7 ¾ inches (13–20 cm)

RINGS
11 + 34 (33–36)

PECTORAL FIN RAYS
17 (16–18)

DORSAL FIN RAYS
18 (16–20)

SPINES
low, rounded bumps only

CORONET
high to very high, rounded (or fluted) top

COLOR/PATTERN
variable pale base color, usually with brown reticulating lines all over, striped snout, dark ring around spines

NOTABLE ANATOMY
very thick rings, narrow body, long snout, double rounded cheek spine, usually prominent eye spine

DEPTH
< 82 feet (25 m)

HABITAT
muddy areas, sponges, sea squirts, man-made holdfasts

CONSERVATION STATUS
Data Deficient

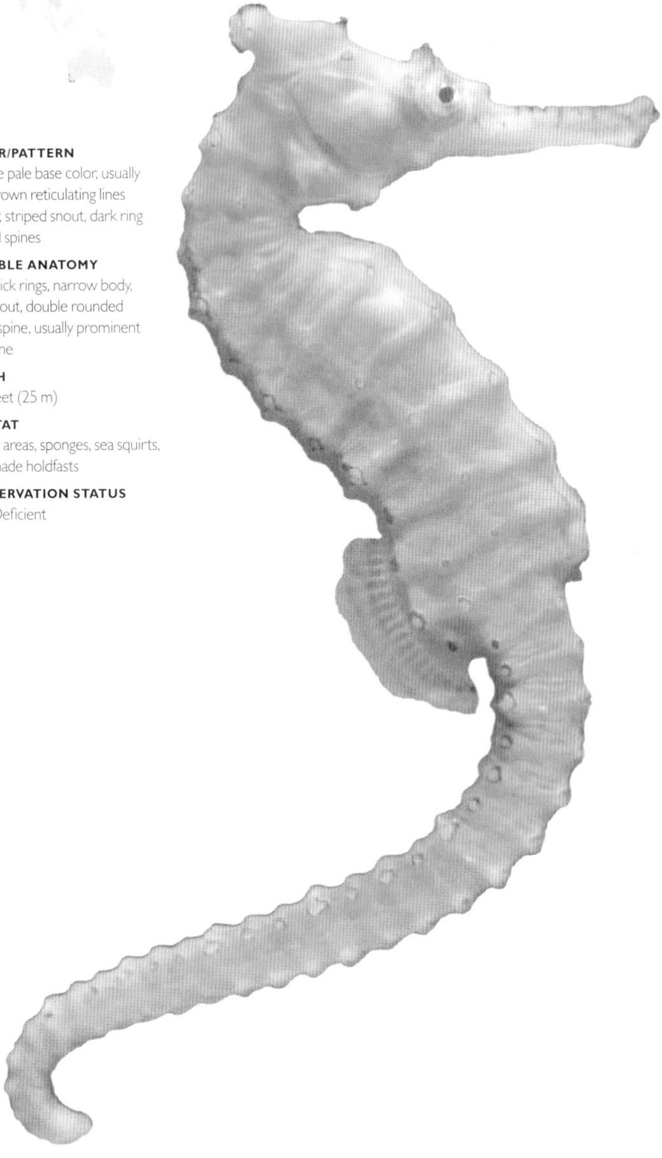

RIGHT The West Australian Seahorse lives in southwest Australia and can be found even relatively far up the Swan River attached to sponges, sea squirts, and man-made objects. It has genetic similarities to *H. angustus* from Shark Bay.

WHITE'S SEAHORSE

ADULT HEIGHT
2½–5¼ inches (6–13 cm)

RINGS
11 + 35 (32–36)

PECTORAL FIN RAYS
16–17 (15–18)

DORSAL FIN RAYS
18 (16–20)

SPINES
variable, low to medium, rounded to sharp

CORONET
high, inclined backward, seven sharp angles or points at tip

COLOR/PATTERN
grayish-brown to yellow, often with reticulating brown lines all over

NOTABLE ANATOMY
long snout, sharp eye spine, usually double cheek spine

DEPTH
< 40 feet (12 m)

HABITAT
seagrass, soft coral, sponges

CONSERVATION STATUS
Data Deficient

White's Seahorse is named in honor of John White, Surgeon General of New South Wales, who first illustrated the seahorse in his *Journal of a Voyage to New South Wales*, published in 1790. At the time he noted that "this animal, like the Flying-fish, being commonly known, a description is not necessary. It is the *Syngnathus hippocampus* of Linnaeus' *Systema Naturae*." Little did he know how complicated seahorse taxonomy actually is.

White's Seahorse is a medium-sized species, with a tall coronet. Like *H. capensis*, it is estuarine with a very restricted distribution: nine estuaries across 187 miles (300 km) of New South Wales, Australia. Each population is self-replenishing with little exchange among populations.

H. whitei was the first seahorse species to be studied extensively in the wild. In the late 1980s Amanda Vincent investigated their social behavior, and documented their courtship and mating. She spent hundreds of hours underwater, patiently observing these fascinating creatures in murky harbor waters, and learning that they maintained faithful pair bonds, and consistent home ranges. She discovered that their courtship lasted from a few minutes to more than an hour, and involved color changes and synchronized promenading across the sea floor, often with tails entwined, and culminated with joint rising in the water column and interlocking of their genitals. She mostly worked in Watson's Bay, Sydney Harbour, because it held a (relatively) large population. However, even here at their densest, only one could be found every 54 square feet (5 m^2).

The restricted range and fragmented population structure of *H. whitei* are both cause for conservation concern. Some of the densest populations are found attached to the swimming nets that protect beaches from large predators such as sharks. Ideally, cleaning the nets should be done in a way that minimizes disturbance to the seahorses, for example leaving the weeds on the bottom 4 feet (1.2 m) of the net, cleaning it in segments, and doing so in winter when the seahorses are not breeding.

RIGHT The tall coronet of White's Seahorse, with its seven spines or angles, is a distinctive feature of this species. It is most commonly found in seagrass beds in southeast Australia attached to sponges or soft corals.

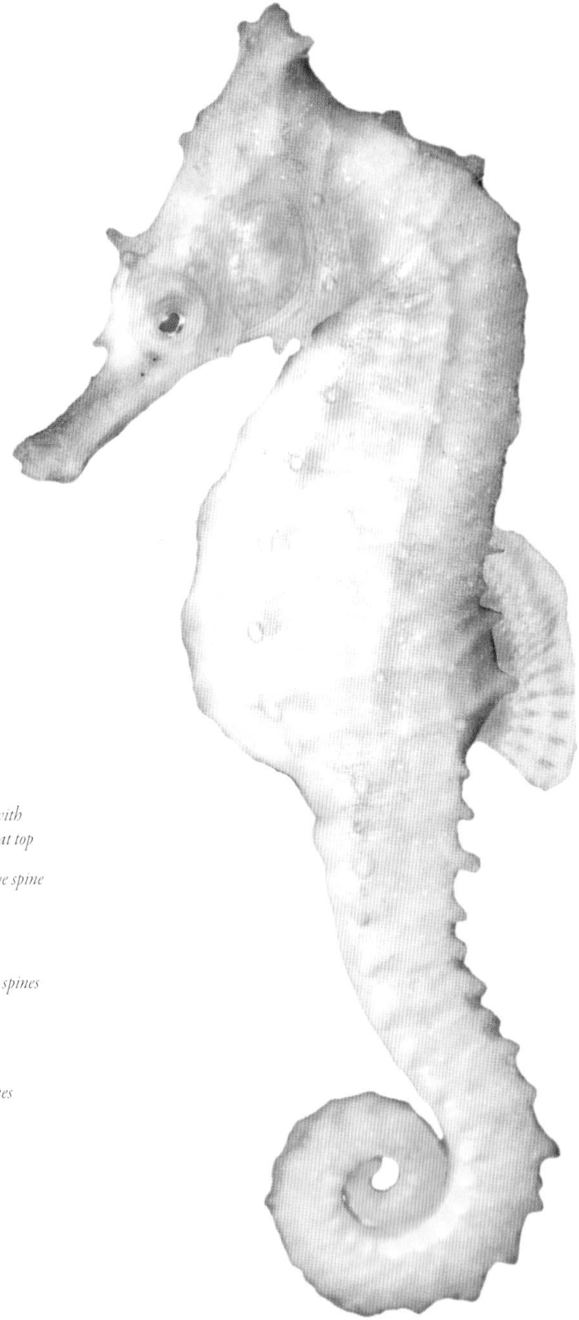

ACTUAL SIZE

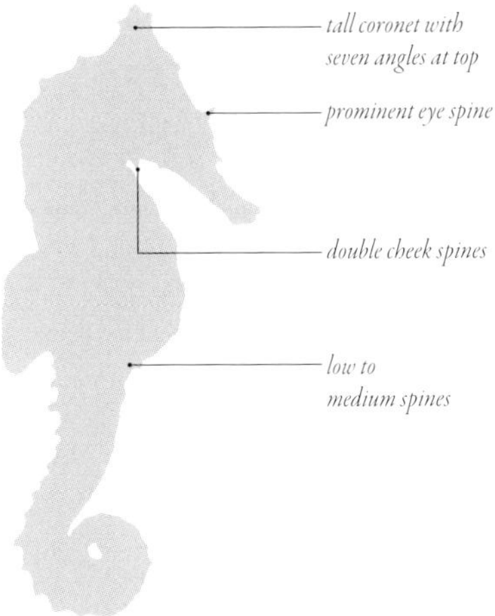

EAST AFRICAN GIRAFFE SEAHORSE

The East African Giraffe Seahorse is relatively uncommon. Like *H. trimaculatus* and *H. planifrons* it sometimes has distinctive black spots on trunk rings 1, 4, and 7; however, unlike these other two species, it has a very tall coronet, which also sometimes has a black spot on its tip. Its tall coronet has led to occasional misidentification resulting in incorrect reporting of *H. whitei* (an Australian species) in east Africa.

Fishers do sometimes catch *H. camelopardalis* accidentally in their nets. Recently, one specimen, apparently of this species, was caught in India. It is unknown whether this represents a native population, or a lone individual that had been carried there on the ocean currents from east Africa.

In Zanzibar, the ashes of a seahorse, mixed with plants, and poured over fishing gear, is apparently recommended to rid the gear of bad spells and attract fish.

ADULT HEIGHT
$2\frac{1}{2}$–4 inches (6–10 cm)

RINGS
11 + 38

PECTORAL FIN RAYS
17–18

DORSAL FIN RAYS
19–22

SPINES
low or rounded, prominent eye spine and pre-coronet spine

CORONET
very high, inclined backward, with a rounded tip

COLOR/PATTERN
variable, dark brown spots on tip of coronet and on trunk rings 1, 4, and 7

NOTABLE ANATOMY
low, or prominent eye spine, short snout

DEPTH
< 148 feet (45 m)

HABITAT
unknown, although it has been photographed in very shallow seagrass

CONSERVATION STATUS
Data Deficient

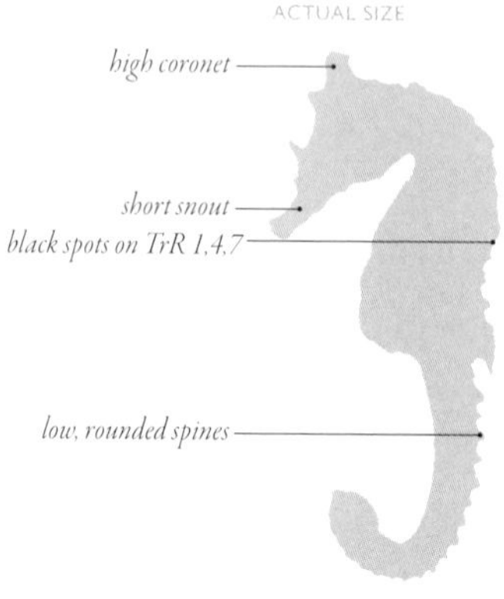

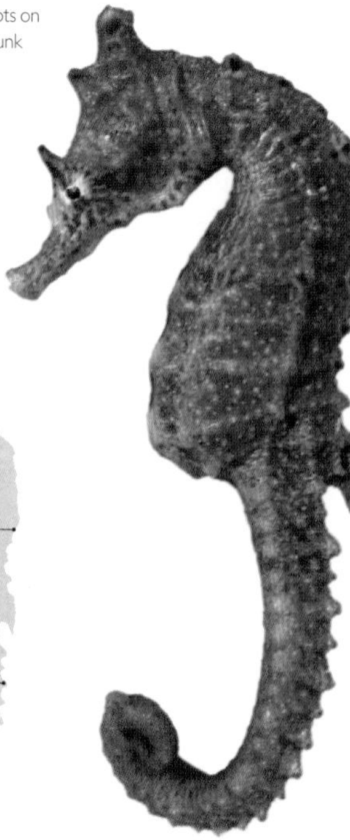

RIGHT Underwater photographs of the East African Giraffe Seahorse show it in seagrass habitats, however little is known about its biology or life history. It is recognized by its tall coronet, and it frequently has black spots on its dorsal trunk ridge and tip of coronet.

FALSE-EYED SEAHORSE

The False-eyed Seahorse has distinctive spots on trunk rings 4 and 7 that appear to be cut into two parts. Genetic research has confirmed that this species, although related to the more widespread Three-spot Seahorse, differs by approximately 3 percent. It is apparently restricted to Shark Bay in Western Australia, and is found in relatively shallow seagrass habitats. Shark Bay is in the most westerly part of Australia, with a marine area of nearly 6,000 square miles (15,500 km^2), an average depth of 30 feet (9 m), and includes one of the largest expanses of seagrass in the world. It is a UNESCO World Heritage Site that straddles the tropical and temperate zones and has an amazing diversity of 12 different species of seagrass. No focused research has been conducted on *H. planifrons* to date.

ADULT HEIGHT
2¾–4 inches (7–10 cm)

RINGS
11 + 39 (39–41)

PECTORAL FIN RAYS
17 (16–18)

DORSAL FIN RAYS
23 (21–23)

SPINES
low or none

CORONET
five tiny points, slightly raised

COLOR/PATTERN
gray-brown, dark keel, "split" spots on TrR 4 and 7, some individuals have "zebra-striped" pattern

NOTABLE ANATOMY
short snout, "split" spots, sharp, recurved cheek and eye spines

DEPTH
< 33 feet (10 m)

HABITAT
seagrass

CONSERVATION STATUS
Vulnerable

ACTUAL SIZE

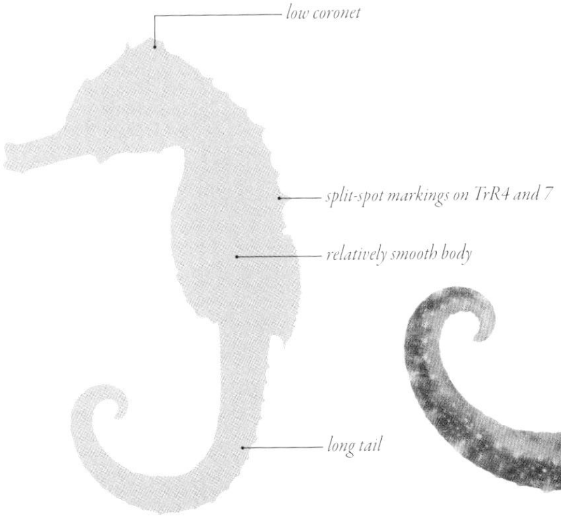

LEFT The False-eyed Seahorse is found only in Shark Bay in Western Australia. It has characteristic split-spot markings on TrR4 and 7 (that look like eyes, although they are not very dark in this photo). They live in shallow water among seagrass.

THREE-SPOT SEAHORSE

ADULT HEIGHT
3¼–6¾ inches (8–17 cm)

RINGS
11 + 40–41 (38–43)

PECTORAL FIN RAYS
17–18 (16–19)

DORSAL FIN RAYS
20 (18–22)

SPINES
low or slightly raised

CORONET
very low, five tiny points

COLOR/PATTERN
golden orange, sandy, or completely black, large dark spots on trunk rings 1, 4, and 7, some individuals have "zebra-striped" pattern

NOTABLE ANATOMY
sharp, hook-like cheek and eye spines

DEPTH
33–131 feet (10–40 m)

HABITAT
sand, gravel

CONSERVATION STATUS
Vulnerable

The Three-spot Seahorse is so named because of the dark spots on the dorsolateral surface of trunk rings 1, 4, and 7 (*tri* = three, *maculatus* = spot). However, not all specimens have these distinctive spots. Specimens with spots tend to be males, while those without spots tend to be females, suggesting that the trait may be gender-linked. Most individuals are brown or mottled in color, although some are black and white striped. These "zebra-striped" specimens are frequently misidentified as *H. zebra*, and it is tempting to consider them a different species. However, genetic studies (sequencing part of the mitochondrial DNA) show no significant difference between the types.

All *H. trimaculatus* specimens have distinctive sharp, hook-like spines on either side of their throat, and above their eyes, and a very small coronet, barely more than five small sharp points raised above the ridge of the head. They also have very long tails.

The Three-spot Seahorse is a widespread species, occurring from India to the Philippines, Japan, northeast Australia, and Tahiti. They are commonly caught in trawl fishing gear, and appear to live in sandy, open habitats at depths of 33–131 feet (10–40 m). Genetic studies reveal a distinct break (1.9–2.9 percent in mtDNA sequences) within Southeast Asia. Unlike other marine species, which tend to fall into two groups based on their ocean basin of origin (Indian versus Pacific), the break in *H. trimaculatus* runs roughly north–south, following Wallace's Line (the biogeographic

ACTUAL SIZE

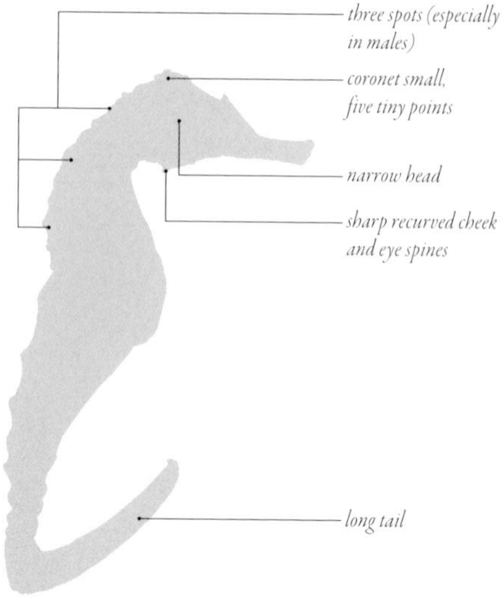

boundary, described by Alfred Russell Wallace in 1845), that separates terrestrial Asian (west) and Australasian (east) plants and animals.

Being a "smooth" species without spines, it is highly sought after for traditional Chinese medicine, and the large volumes that come up as by-catch from trawl fisheries, and from there enter the dried trade, make this one of the most highly traded species. Along with *H. algiricus*, *H. barbouri*, and *H. histrix*, this species has recently been subjected to a "Significant Trade Review" through CITES.

BELOW As seen in this photograph, the Three-spot Seahorse usually has three distinctive black marks on its back. It also has distinctive sharp, recurved cheek and eye spines and a narrow head. Males commonly have a dark ventral mid-line. Some specimens are "zebra-striped."

CROWNED SEAHORSE

The Crowned Seahorse can be identified by its enormous coronet, which can be as tall as one and half times the length of its snout. It also has distinctive flattened "wing-like" spines on either side of its short dorsal fin base, and other irregular thin, round-tipped spines.

The name *Hippocampus coronatus* has frequently been misapplied to *Hippocampus sindonis*, although morphological and genetic data confirm that they are two separate species.

The distribution of the Crowned Seahorse is restricted to Japan and South Korea, where it is found in seagrass beds along with *Hippocampus mohnikei* at densities of ~1 individual per 3,768 square feet (350 m^2). Crowned Seahorses are reproductively mature at about 2½ inches (6 cm) in height and breeding occurs between July and November. Brood sizes are typically between 10 and 50.

ADULT HEIGHT
2¼–5¼ inches (6–13 cm)

RINGS
10 + 39 (38–40)

PECTORAL FIN RAYS
12

DORSAL FIN RAYS
14

SPINES
very irregular, most body angles have none, some are long, thin, blunt-tipped

CORONET
extremely tall, tip fluted, turned back

COLOR/PATTERN
yellowish, marbled or dotted with dark brown

NOTABLE ANATOMY
very tall coronet, short dorsal fin base flanked by "wing-like" flattened projections

DEPTH
< 33 feet (10 m)

HABITAT
seagrass, floating algae

CONSERVATION STATUS
Data Deficient

ACTUAL SIZE

RIGHT The very tall coronet, and distinctive enlarged and flattened spines on the body of the Crowned Seahorse immediately identify it. It is found in Japan and South Korea.

JAPANESE SEAHORSE

Although *H. mohnikei* is called the Japanese Seahorse, it actually has a wider geographic distribution than the other two Japanese miniatures, being found in Korea, China, Vietnam, Cambodia, Thailand, and India. That said, there is some evidence to suggest that even though these populations look morphologically the same, they may represent cryptic species that are genetically quite dissimilar. Further research is needed.

What we do know is that the Japanese Seahorse is a relatively small seahorse with a long tail, a very short snout, wedge-like coronet, and double rounded cheek spines. It is usually black, dark brown or yellow, sometimes with white bands on its tail. It can be found in shallow seagrass beds during the summer months and may move to deeper water when the seagrass dies off in the winter. It breeds from May to September. Newborns have a vestigial caudal fin that is lost within a few days of birth. Juveniles spend at least a month in the plankton before settling to the bottom.

ADULT HEIGHT
2–3¼ inches (5–8 cm)

RINGS
11 + 38 (37–40)

PECTORAL FIN RAYS
13 (12–14)

DORSAL FIN RAYS
15–16

SPINES
low

CORONET
low, wedge-like crest

COLOR/PATTERN
usually black or dark brown, or yellow, may have striped tail

NOTABLE ANATOMY
very long tail, double rounded cheek and eye spines

DEPTH
< 33 feet (10 m), pelagic young offshore

HABITAT
seagrass

CONSERVATION STATUS
Data Deficient

BELOW The Japanese Seahorse is found from India to Japan in shallow sandy seagrass beds. It is relatively smooth with wedge-like coronet, but a very long tail and double, low cheek spines.

ACTUAL SIZE

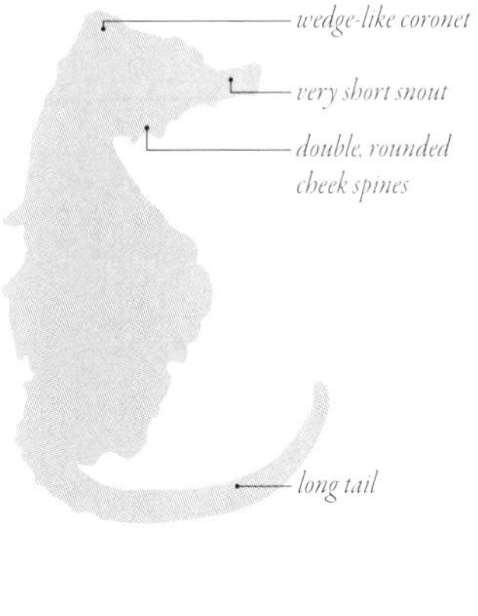

Hippocampus sindonis
JORDAN & SNYDER, 1902

SINDO'S SEAHORSE

Sindo's Seahorse is a small, colorful, seahorse with an angular coronet that usually has branched skin filaments on its head and back, and a double spine above each eye. A particularly long filament, commonly attached to the front of the coronet, has led to confusion with *H. coronatus*, but the two are separate species. They both have ten trunk rings (as opposed to the more common 11) and it is likely that they are closely related. They are also both only found in Japan and the southern Korean peninsula.

Very little is known about this seahorse. It was first discovered by scientists aboard the US Fish Commission steamer, *Albatross*, during its 1899–1900 expedition to Japan and the South Pacific. It was dredged from a depth of 187–246 feet (57–75 m) in Totomi Bay off Hamamatsu. It was named for Michitaro Sindo, assistant curator of fishes at Stanford University.

ADULT HEIGHT
1½–3¼ inches (4–8 cm)

RINGS
10 + 37 (36–38)

PECTORAL FIN RAYS
12–14

DORSAL FIN RAYS
12 (11–15)

SPINES
well-developed, blunt-tipped

CORONET
medium height, well-developed, angular

COLOR/PATTERN
greenish-gray, red, yellow, orange, mottled

NOTABLE ANATOMY
usually long skin filament at front of coronet

DEPTH
<164–246 feet (50–75 m)

HABITAT
volcanic mud, sand

CONSERVATION STATUS
Least Concern

BELOW Sindo's Seahorse is a small Japanese species with a squarish coronet and ten trunk rings. It often had skin filaments attached to its head and back and comes in a wide range of colors. This photograph was taken in Mie Prefecture, Japan.

ACTUAL SIZE

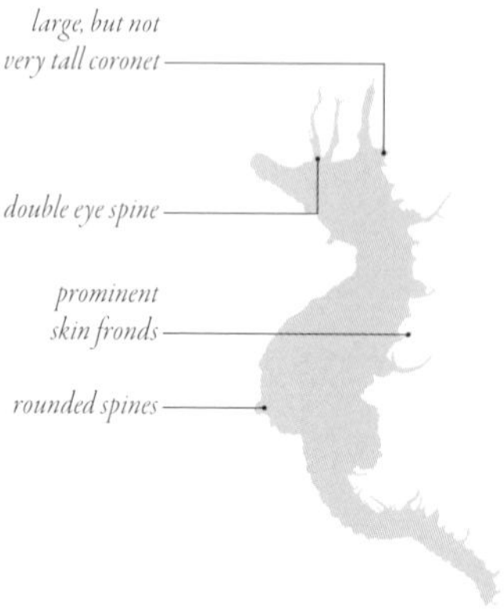

Hippocampus borboniensis
DUMÉRIL, 1870

RÉUNION SEAHORSE

The Réunion Seahorse was first described based on a specimen from the island of Réunion. It has similarities to members of the *H. kuda* complex, however, it is more knobbly, and further research may show that it is actually aligned with the "semi-spiny *kuda* relatives" rather than the smoother members of the "*kuda*-complex."

It is found on the islands of Mauritius and Madagascar, and may also occur off the east coast of Africa, although identifications of this species from there are tentative.

Nothing is known of its ecology or behavior in the wild, although photos that have been identified as this species show it on sand at a depth of 40 feet (12 m).

ADULT HEIGHT
3¼–5½ inches (8–14 cm)

RINGS
11 + 35–36 (34–38)

PECTORAL FIN RAYS
15–16

DORSAL FIN RAYS
17 (16–18)

SPINES
well-developed rounded knobs

CORONET
low, with five rounded knobs

COLOR/PATTERN
dusty green-brown with dusty yellow dots, marbling and broken lines on head, or uniformly dark

NOTABLE ANATOMY
usually with prominent rounded eye spines

DEPTH
< 40 feet (12 m)

HABITAT
sand, seagrass

CONSERVATION STATUS
Data Deficient

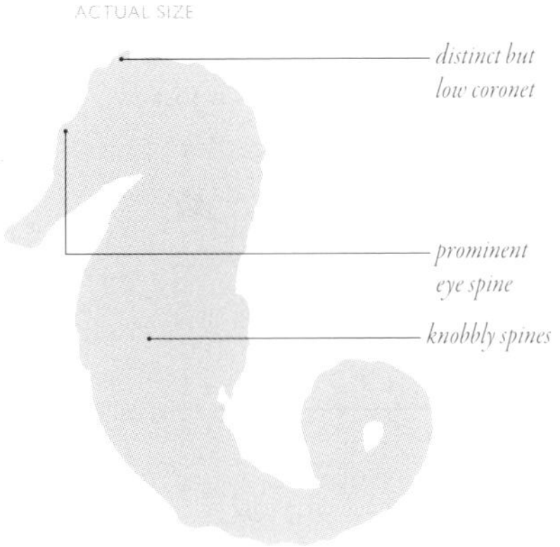

LEFT This photograph is tentatively identified as the Réunion Seahorse although it is less knobbly that the type specimen. The photograph was taken in Réunion.

GREAT SEAHORSE

The Great Seahorse is one of the largest seahorse species, growing to nearly 11¾ inches (30 cm) in height. Because of its large size and smooth appearance (when adult), it is probably the most highly favored species for traditional Chinese medicine. It has a narrow body, and young specimens may have slightly better developed spines than adults. This can sometimes lead to identification confusions with *H. spinosissimus*.

This species is most commonly found in deep water. It is rarely photographed in its natural habitat, but is frequently accidentally caught by trawl operations in Southeast Asia, and thus enters the dried-seahorse trade.

The Great Seahorse has been recorded from eastern Australia, and some specimens labeled *H. tristis* by Rudie Kuiter are likely to be this species as well. It is unclear whether specimens from the Red Sea and Oman Sea (sometimes referred to as *H. suezensis*) are part of this same species. Further research is needed.

LEFT The Great Seahorse is one of the largest species and is usually found at depths of more than 66 feet (20 m). It has a narrow body, fine white vertical lines, a relatively long snout and low rounded angles to its body ridges. It has long "drooping" cheek spines.

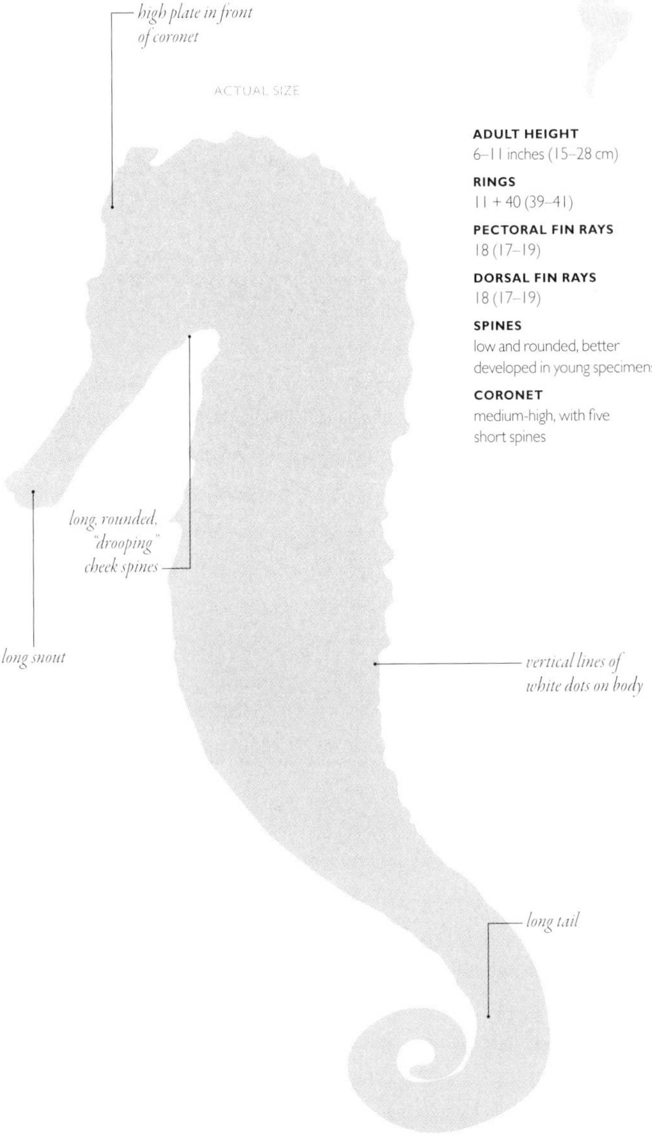

ADULT HEIGHT
6–11 inches (15–28 cm)

RINGS
11 + 40 (39–41)

PECTORAL FIN RAYS
18 (17–19)

DORSAL FIN RAYS
18 (17–19)

SPINES
low and rounded, better developed in young specimens

CORONET
medium-high, with five short spines

COLOR/PATTERN
uniform pale, orange or red, often with vertical lines of tiny white dots

NOTABLE ANATOMY
long, slightly backward-pointing, rounded cheek spines, high plate in front of coronet

DEPTH
66–492 feet (20–150 m)

HABITAT
soft bottom, gorgonian corals and sea whips

CONSERVATION STATUS
Vulnerable

Hippocampus spinosissimus

WEBER, 1913

HEDGEHOG SEAHORSE

ADULT HEIGHT
2–$6\frac{3}{4}$ inches (5–17 cm)

RINGS
11 + 36 (33–39)

PECTORAL FIN RAYS
17 (16–19)

DORSAL FIN RAYS
17–18 (16–20)

SPINES
well-developed, sharp or blunt-tipped, usually longer on TrR1, 4, 7, 11 and in regular series on tail

CORONET
low to medium with four to five sharp spines

COLOR/PATTERN
red, orange, brown, black, yellow, white, may be speckled and often with distinctive "saddle-like" markings

NOTABLE ANATOMY
single or double cheek spines, males with prominent spines bordering pouch, no (or very small) nose spine, no (or very small) spine in front of coronet

DEPTH
33–230 feet (10–70 m)

HABITAT
sand, soft corals, octocorals, macroalgae, urchins

CONSERVATION STATUS
Vulnerable

The Hedgehog Seahorse is another spiny seahorse that has frequently been confused with *H. histrix*. It can be distinguished from the latter based on its shorter snout, shorter spines, and more tail rings. It was first described on the basis of two specimens that were dredged from a depth of 226 feet (69 m) in the Sapeh Strait, between Sumbawa and Komodo in Indonesia. Both specimens were small (less than $2\frac{1}{2}$ inches/6 cm) but were clearly mature, since they had well-developed pouches (and one was even pregnant). Most examples of this species are larger, although there is some disagreement as to whether or not they are indeed the same.

Assuming that they are (pending further research), the Hedgehog Seahorse is widely distributed throughout Southeast Asia and northern Australia. Specimens from Queensland were described as a new species in 2001 (*H. queenslandicus*), but subsequent genetic data do not separate them from Asian specimens. Other names that have been used for this species include *H. alatus, H. arnei, H. polytaenia, H. moluccensis,* and *H. semispinosus*. Part of the reason there are so many different names is because the color of this species varies widely: they may be red, orange, black, brown, yellow, or white. Specimens often have distinctive "saddle-like" markings at intervals over their trunk and tail. Some individuals are speckled either with white (darker individuals) or red/brown (lighter individuals) spots. There are also variation in the form of the cheek spines (single or double), the number of spines on the coronet (four or five), and at least three different (but not drastically different) genetic types. No clear pattern in these variations has yet been elucidated, however they warrant further investigation.

The Hedgehog Seahorse lives in fairly deep water, usually > 33 feet (10 m), although is occasionally found at shallower depths. It occurs in relatively open habitats. It has been observed using soft corals, octocorals, and even mobile pencil urchins as holdfasts. It is a common by-catch species of trawlers operating in the South China Sea, and frequently ends up in the traditional medicine trade.

RIGHT The Hedgehog Seahorse has been given a number of different names and it is quite variable in color and pattern. Some are plain, and some are mottled, whereas others (like this one) have distinctive "saddle-like" markings. Its snout is medium-long, as are its spines, and it has a distinctive spiny coronet.

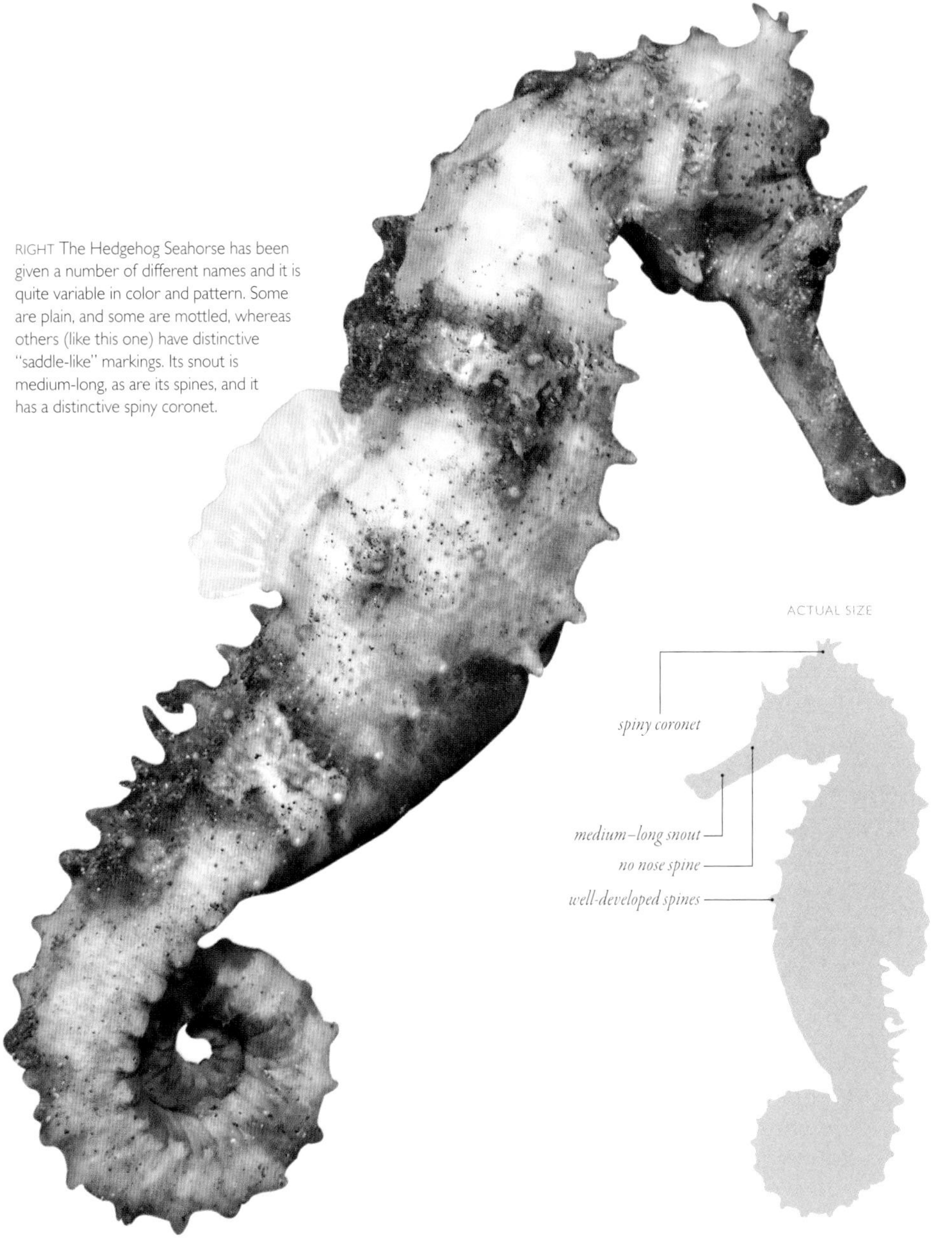

WEST AFRICAN SEAHORSE

The West African Seahorse was originally described from a specimen sent from Algiers (hence the name) to the Paris Museum by a M. Guichot. Since then, there have been no additional sightings in Algeria, and all other specimens have come from West Africa. The distribution of *H. algiricus* is now considered to be from Senegal to Angola.

Genetically speaking, the West African Seahorse is extremely closely related to the Slender Seahorse (*H. reidi*) from Brazil and the Caribbean, and it is likely that the common ancestor of the sub-group that contains both these species colonized the Atlantic from the Indo-Pacific around the tip of South Africa.

Little is known about the life history of the West African Seahorse, although it seems to be more common in southern Senegal, The Gambia, Guinea Bissau, and Guinea Conakry than in northern Senegal. This is likely due to the presence of more extensive shallow habitats in the south.

The first known trade in West African Seahorses began in the 1990s, but has increased rapidly in the past 10–15 years as supply has begun to dwindle in other parts of the world. Most seahorses are incidentally caught by artisanal fishers, but small numbers from many fishers add up significantly. In 2010 half a million seahorses were recorded as being imported to Hong Kong SAR, China, and Taiwan from Senegal and Guinea precipitating a CITES Review of Significant Trade (RST) in 2012.

ADULT HEIGHT
3½–7½ inches (9–19 cm)

RINGS
11 + 36 (35–37)

PECTORAL FIN RAYS
16–17

DORSAL FIN RAYS
17–18

SPINES
low, rounded bumps only

CORONET
relatively low, rounded, and overhanging at the back, flat topped, or with a slight depression in center

COLOR/PATTERN
usually black or brown, but yellow, red, green, or orange also reported

NOTABLE ANATOMY
broad, almost double cheek and eye spines, thick body rings

DEPTH
< 164 feet (50 m), mostly shallow

HABITAT
silt, soft bottom, pebbles, sand, seagrass, seaweed, estuaries

CONSERVATION STATUS
Vulnerable

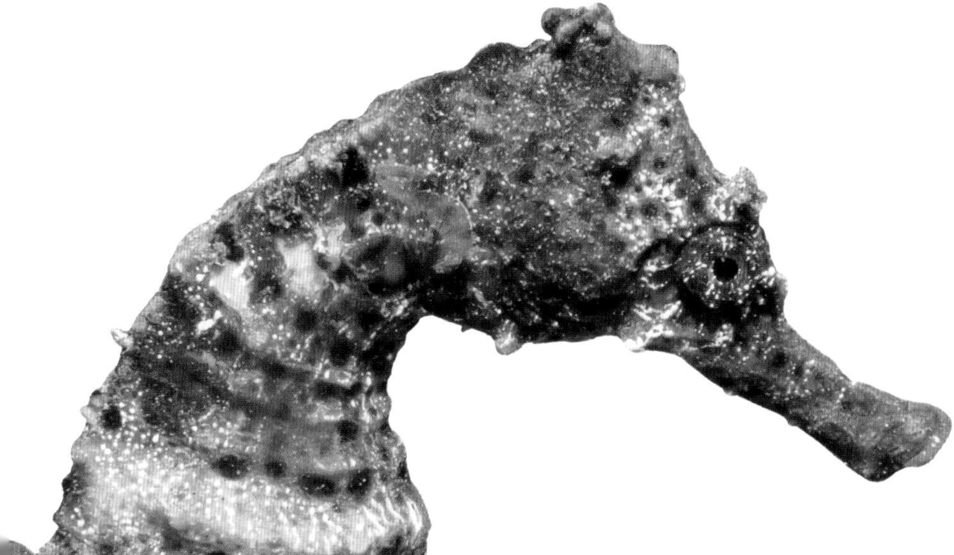

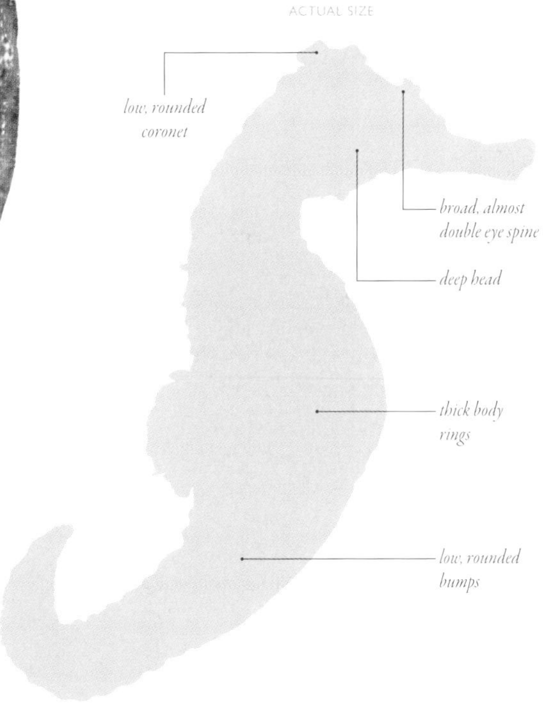

ABOVE The West African Seahorse is a smooth species, with a rounded coronet. It lives in fairly shallow water in sandy and muddy habitats, especially estuaries. It is most common in southern Senegal, The Gambia, Guinea Bissau, and Guinea Conakry, and is very closely related to the Slender Seahorse from the western Atlantic.

KNYSNA SEAHORSE

The Knysna Seahorse is the only fully estuarine species of seahorse, and can withstand transfer from seawater (with a salt concentration of 35 parts per thousand) to almost pure freshwater (one part per thousand).

It is also the species that has the smallest distribution—its entire area of occupancy is less than 20 square miles (50 km^2). Despite extensive searches, it is known to inhabit only three estuaries in South Africa and has the dubious distinction of being the only seahorse that is listed as Endangered on the IUCN Red List of Threatened Species. It is closely related to other members of the *H. kuda* complex, although it is small and has a distinctive morphology with a smooth body, short snout, rounded head and no coronet.

Genetic studies and field surveys have shown that the *H. capensis* population in Knysna Estuary is the largest, most diverse, and most stable. The population in Swartvlei suffered a mass mortality in 1991 (more than 3,000 dead seahorses were found after a breach of the estuary mouth), and decreased in size by 80 percent between 2002–2003, highlighting the vulnerability of this extremely geographically restricted species.

ADULT HEIGHT
2–4¾ inches (5–12 cm)

RINGS
11 + 34 (32–37)

PECTORAL FIN RAYS
15 (14–17)

DORSAL FIN RAYS
17 (16–18)

SPINES
none on body, very short and blunt on tail

CORONET
none

COLOR/PATTERN
usually mottled brown with darker patches but may be white, yellow, orange, beige, green, black

NOTABLE ANATOMY
short snout, males with slight keel

DEPTH
< 66 feet (20 m), mostly shallow

HABITAT
estuaries

CONSERVATION STATUS
Endangered

Based on a small mark-recapture study, there seems to be some site fidelity among individuals although not as pronounced as other species such as *H. comes*. There were also no obvious social groupings, greetings, or courtship observed. Individuals reach sexual maturity in less than one year and can live for at least three years in captivity. Longevity in the wild is unknown.

RIGHT The Knsyna Seahorse is restricted to just three estuaries in South Africa. It is part of the Kuda-complex of seahorses, however it looks quite distinctive with a short snout, smooth body, and very rounded head without a perceptible coronet.

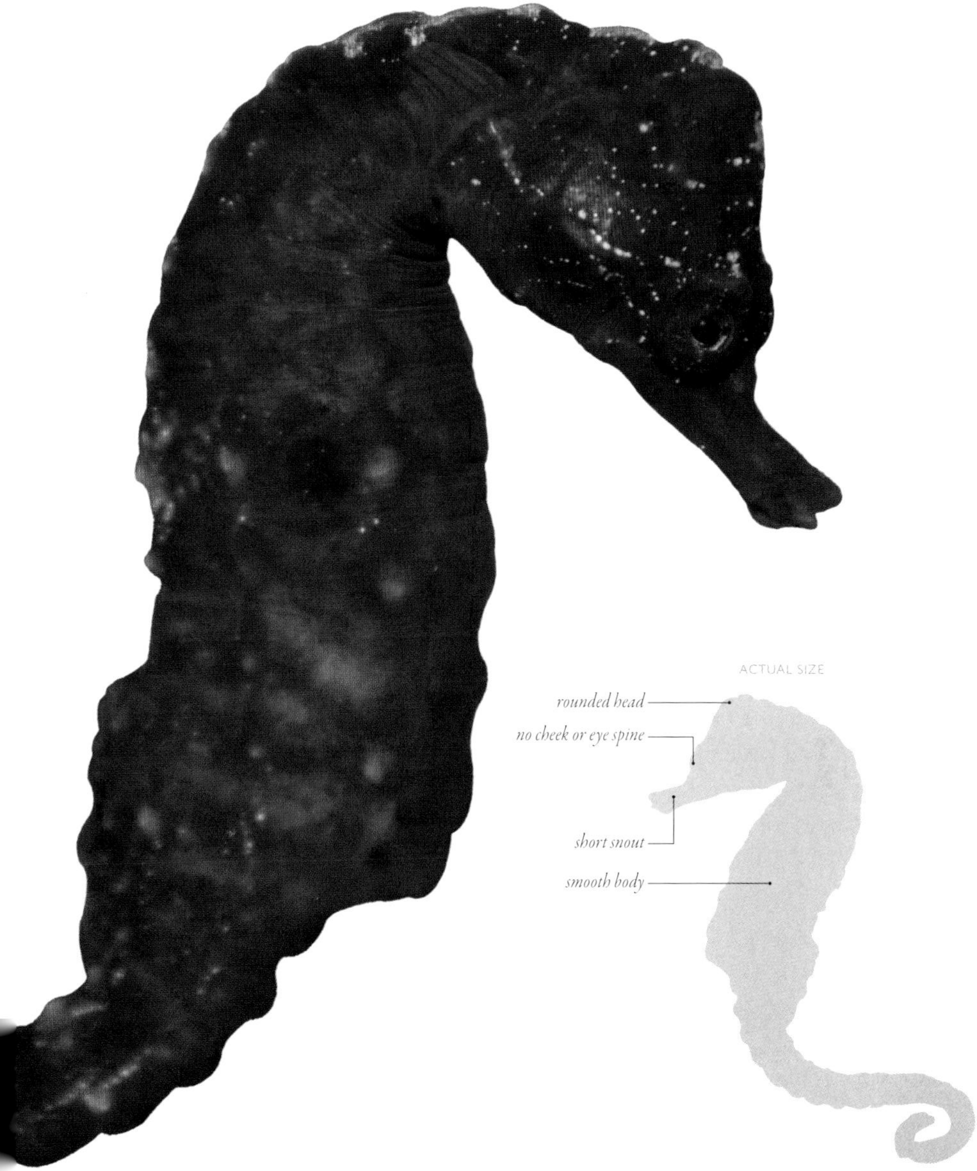
ACTUAL SIZE
rounded head
no cheek or eye spine
short snout
smooth body

FISHER'S SEAHORSE

Fisher's Seahorse is found only in Hawaii, and is sometimes called the Hawa'ii Seahorse. This species is part of the *H. kuda* complex, although it does not look like the others. It may have evolved from *H. ingens* or *H. kuda* after a long-distance colonization event, and subsequent isolation in the middle of the Pacific.

Genetic data from the Barcode of Life Project also group Fisher's Seahorse with the *H. kuda* clade, but suggest that it is 3–4 percent different from other members of the group. Further research is required to clarify its taxonomic position.

Little is known of the life history or behavior of Fisher's Seahorse, although one was recently brought up from French Frigate Shoals, attached to a crab pot at 328 feet (100 m) depth, and others were obtained from the stomachs of large pelagic fish (for example, trevally or dolphin fish). A very recent specimen photographed *in situ* in the Northwestern Hawaiian Islands at a depth of 286–289 feet (87–88 m) may be an example of this species.

ADULT HEIGHT
2–3¼ inches (5–8 cm)

RINGS
11 + 37–38 (36–39)

PECTORAL FIN RAYS
15 (13–16)

DORSAL FIN RAYS
17–18

SPINES
small, but quite sharp, occasional spines expanded and flattened

CORONET
slightly raised with five tiny points

COLOR/PATTERN
golden orange, red, pink, brown

NOTABLE ANATOMY
small, sharp, slightly hooked double eye and cheek spines, sharp spine in front of coronet and two behind coronet

DEPTH
< 328 feet (100 m)

HABITAT
unknown

CONSERVATION STATUS
Data Deficient

ACTUAL SIZE

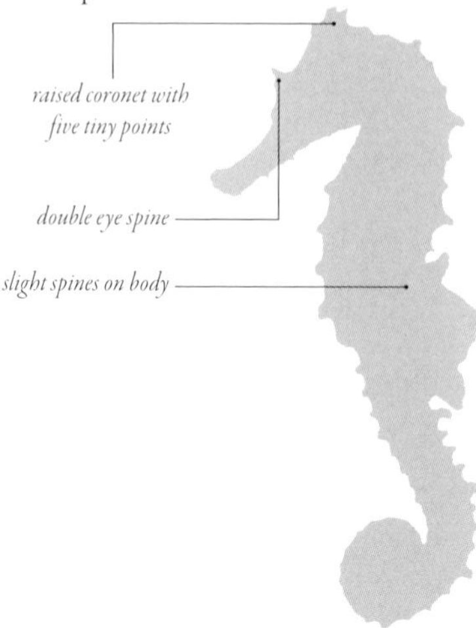

RIGHT Fisher's Seahorse is a relatively small species that can be found at great depths. It has a narrow head and a small coronet, and is believed to come to the surface at night. Its cheek and eye spines are slightly curved backwards and pointed (like the Three-spot Seahorse).

SEA PONY

The Sea Pony is another small member of the *H. kuda* clade, described originally from the Red Sea. It is distinguished from the larger *H. kuda* by a shorter snout, fewer tail rings, and a more crest-like head without a clearly defined coronet. Specimens from India and East Africa with short snouts and lacking a distinct coronet have been called *H. fuscus*, but further research is needed to determine whether or not they are the same as the ones from the Red Sea.

In captivity, Sea Ponies form stable pairs and are sexually monogamous. Their maximum brood size is approximately 150 young, and the young reach reproductive maturity and can themselves give birth in as little as four months.

ADULT HEIGHT
3¼–4¾ inches (8–12 cm)

RINGS
11 + 34 (33–37)

PECTORAL FIN RAYS
15 (14–16)

DORSAL FIN RAYS
16 (14–17)

SPINES
low, body relatively smooth

CORONET
none, rough crest only

COLOR/PATTERN
usually dark, but can be yellow or mottled brown

NOTABLE ANATOMY
short snout, head relatively large and deep

DEPTH
< 33 feet (10 m)

HABITAT
seagrass, algal beds, gravel, protected bays

CONSERVATION STATUS
Data Deficient

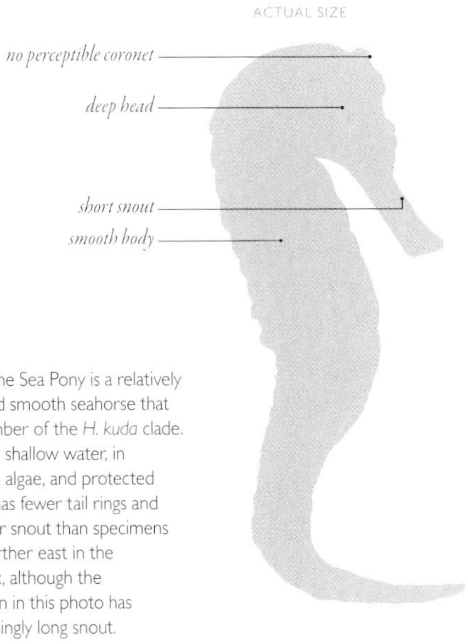

RIGHT The Sea Pony is a relatively small and smooth seahorse that is a member of the *H. kuda* clade. It lives in shallow water, in seagrass, algae, and protected bays. It has fewer tail rings and a shorter snout than specimens from further east in the complex, although the specimen in this photo has a surprisingly long snout.

PACIFIC SEAHORSE

The Pacific Seahorse is one of the largest seahorse species in the world. It can grow to over a foot (30 cm) in length, and is the only species of seahorse to be found on the west coast of the American continent. Its sister (i.e. most closely related) species is the Slender Seahorse (*H. reidi*), found in the Caribbean, on the other side of the Isthmus of Panama. The latest time that these two shared a common ancestor cannot have been less than 3.1–3.5 million years ago when the land bridge between North and South America formed, permanently separating the Pacific and Atlantic Oceans. Using this date, and the genetic divergence between these two seahorses, it is possible to "calibrate" a molecular clock (assuming a relatively constant rate of molecular divergence). This rate of evolution can then be used to determine how long ago other pairs of seahorses diverged.

H. ingens specimens from Peru to San Diego are genetically very similar to one another. This suggests either high dispersal among the populations, or very recent expansion of the populations following a bottleneck in population size. The latter is possible since seahorses in the northern part of the range were apparently absent for about 100 years between the 1850s and the 1960s, and only recolonized the area recently as a result of increasing sea temperatures. Specimens from the Gulf of California are slightly divergent from those on the Pacific coast, suggesting that they should be treated as a separate management unit when designing conservation strategies.

ADULT HEIGHT
5¼–12¼ inches (13–31 cm)

RINGS
11 + 39 (38–40)

PECTORAL FIN RAYS
16 (15–17)

DORSAL FIN RAYS
19 (18–21)

SPINES
variable, low and rounded, to well-developed with blunt tips

CORONET
medium-high, tilted backward, with five well-defined points, sharp edges, or flanges at top

COLOR/PATTERN
reddish, maroon, brown, gray, yellow, gold, or black, often with fine white lines running vertically down body

NOTABLE ANATOMY
prominent, long, "drooping" cheek spines; prominent, rounded eye spine

DEPTH
usually < 66 feet (20 m), maximum recorded 197 feet (60 m)

HABITAT
seagrass, coral reef, offshore

CONSERVATION STATUS
Vulnerable

Being a large and smooth seahorse, the Pacific Seahorse is highly valued for traditional medicine, and many specimens are caught to satisfy the growing Asian market in California. It is also one of the few species for which the Minimum Size Limit (see Conservation, p. 57), of 4 inches (10 cm) in height, which was put in place as an interim conservation measure, does not serve well. Because it is a large species, it is likely that even at 4 inches (10 cm) in height most specimens are still not reproductively mature, and will not have had a chance to contribute to the next generation prior to being harvested.

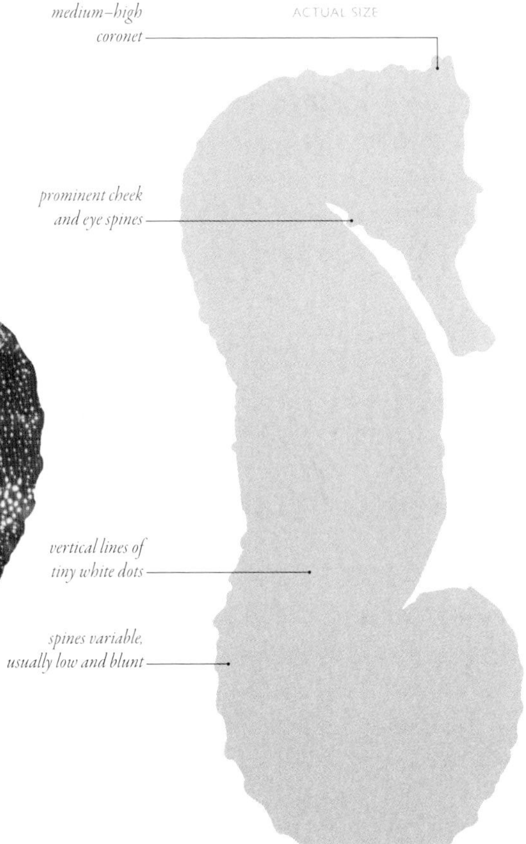

ABOVE The large Pacific Seahorse occurs in a wide variety of colors. It has a medium-high, slightly tipped back coronet, and rounded cheek and eye spines. It usually has vertical lines of tiny white dots on its body and tail. This specimen was photographed in the Galapagos Islands.

Hippocampus kuda

BLEEKER, 1852

COMMON, YELLOW, OR SPOTTED SEAHORSE

ADULT HEIGHT
2¾–6¾ inches (7–17 cm)

RINGS
11 + 36 (34–38)

PECTORAL FIN RAYS
16 (15–18)

DORSAL FIN RAYS
17–18

SPINES
low, rounded bumps only

CORONET
low-medium, rounded, overhanging at the back, often with cup-like depression in top

COLOR/PATTERN
black (often with grainy texture, or fine vertical lines of white dots), pale yellow, or cream (often with brown spots)

NOTABLE ANATOMY
deep head, deep body, thick snout

DEPTH
usually < 33 feet (10 m), but recorded < 164 feet (50 m)

HABITAT
mangroves, seagrass, estuaries

CONSERVATION STATUS
Vulnerable

The name *H. kuda* (Common, Yellow, or Spotted Seahorse), which derives from the Malay word for "horse" (*kuda*), has been used for pretty much any smooth seahorse species in the Indo-Pacific. Bleeker's original description (his first of ten seahorse descriptions) referred to two seahorses from Singapore. However, the whereabouts of these precise specimens is unknown. Pieter Bleeker, a medical doctor and extremely prolific ichthyologist, stationed in Batavia (now Jakarta) from 1842–1860, subsequently used the name for specimens from elsewhere in the East Indian Archipelago. Other species that he described (for example *H. melanospilos* and *H. taeniopterus*), as well as a number described by other authors (*H. novaehebudorum*, *H. taeniops*, *H. hilonis*, *H. aterrimus*, and others) have been synonymized with *H. kuda*. These taxonomic decisions have not met with universal acceptance, however, and further research is warranted. Separate populations of *H. kuda* are often distinct from their neighbors (suggesting limited movement among populations), but overall morphological and genetic variation is slight, and insufficient—in the estimation of Lourie, Pollom, & Foster (in review)—to warrant species distinction.

The Common Seahorse lives in very shallow water, and can often be found in mangroves, seagrass beds, or estuaries. Males are typically black (often with tiny white spots all over), and females are frequently yellow (sometimes with large brown spots).

Choo Chee Kuang, an extremely dedicated young Malaysian researcher, studied *H. kuda* in the Straits of Johor. The seagrass beds where he worked were dredged, polluted, and destroyed by land-reclamation (i.e. sea-filling) and increased boat traffic, resulting in the loss of almost the entire seahorse population. Although Chee Kuang sadly passed away in 2013, his passion and work live on through the SOS (Save Our Seahorses) project that he began in 2004.

Common Seahorses are highly sought after in the traditional medicine trade, and are often bleached to make them more desirable. They are also one of the more common aquarium species and numerous projects, particularly in China, India, and Vietnam have sought to breed this species in captivity.

RIGHT The Common Seahorse is usually either black (as in this picture), or yellow with scattered dark brown spots. Usually the black ones are males, and the yellow ones are females. They live in shallow water, in seagrass beds, muddy or sandy habitats, and estuaries.

ACTUAL SIZE

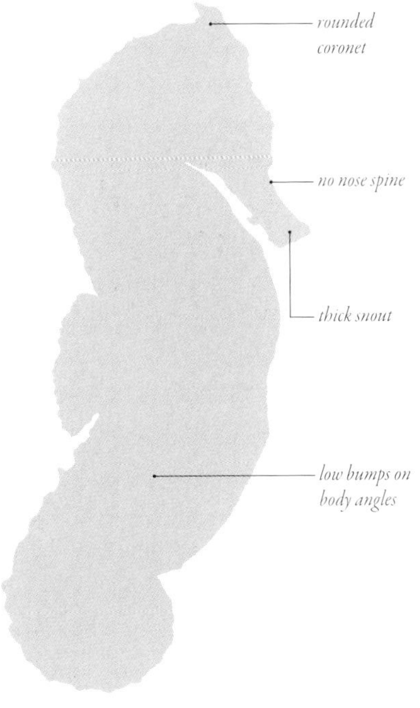

SLENDER SEAHORSE

The Slender Seahorse is the western Atlantic member of the *H. kuda* complex. It is relatively large, and smooth, with a distinctive coronet that looks like a crumpled-up ball of paper. It occurs in a range of colors from brown to yellow, orange, red, and black, and is a highly sought-after aquarium species. It is extremely closely related to the West African Seahorse (*H. algiricus*), and the two may, in fact, be able to interbreed if the opportunity arose. The Slender Seahorse is found in the Caribbean, and on both sides of the Amazon outflow in Brazil, although there is unlikely any ongoing gene flow between these populations.

In a survey in Brazil the population density was ~1 per 431 square feet (40 m^2), and repeated observations of the same individuals in the same locations suggested some site fidelity. Most seahorses were observed alone, however male-female pairs were also common. They were most commonly found in estuaries, and their favored holdfasts seemed to be mangrove roots and fallen branches. Another study showed that home ranges varied from 65–215 square feet (6–20 m^2). There were more females than males in the population, and males had smaller home ranges. The highest densities were in the shallowest areas (< 3 feet (1 m) deep).

Brood sizes of the Slender Seahorse can be up to 1,500, making it one of the most fecund seahorse species. It reaches maturity after only about two months, and in captivity it can produce its first offspring in as little as 81 days.

ADULT HEIGHT
4–7 inches (10–18 cm)

RINGS
11 + 35 (31–39)

PECTORAL FIN RAYS
16 (15–17)

DORSAL FIN RAYS
17 (16–19)

SPINES
none to low, rounded tubercles

CORONET
low–medium, rounded, may be quite large and convoluted (like crumpled piece of paper)

COLOR/PATTERN
red, yellow, orange, black, brown, often profusely spotted, may have paler "saddles" across dorsolateral surface

NOTABLE ANATOMY
broad, almost double cheek and eye spines; long, thick snout

DEPTH
usually shallow, but recorded < 181 feet (55 m)

HABITAT
rocky shore, mangroves, seagrass, macroalgae, artificial structures, estuaries, bivalve beds, sponges

CONSERVATION STATUS
Data Deficient

The life cycle of *H. reidi* has been closed in captivity, allowing for the sale of purely captive-bred seahorses, however specimens are also caught by hand, by breath-hold divers, in waters no deeper than 23 feet (7 m) in Brazil (one of the largest aquarium-exporting countries). The Slender Seahorse is also used for traditional magico-religious purposes in Brazil.

Newborn Slender Seahorses have a vestigial caudal tail fin (with a single fin ray) that is lost within a few days of birth.

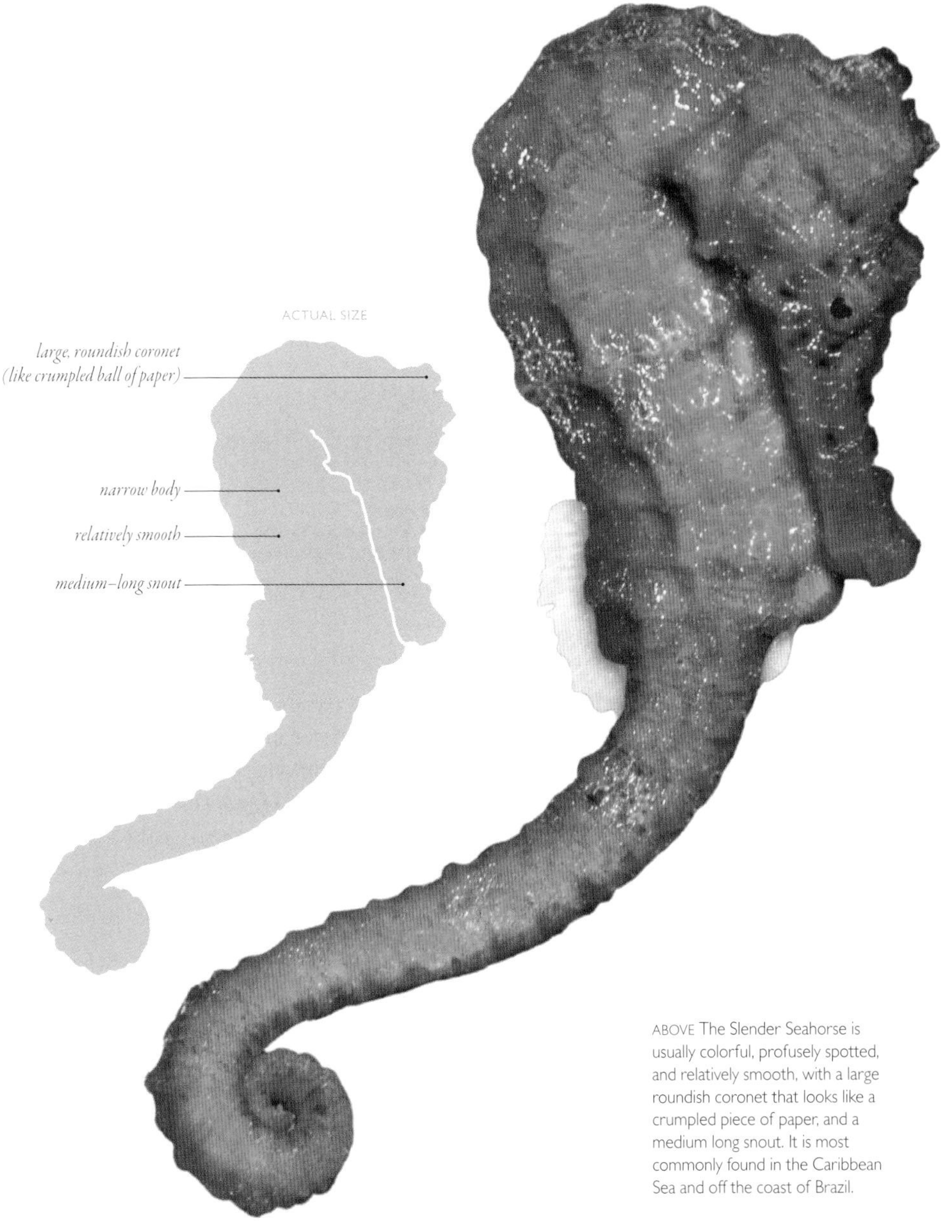

ABOVE The Slender Seahorse is usually colorful, profusely spotted, and relatively smooth, with a large roundish coronet that looks like a crumpled piece of paper, and a medium long snout. It is most commonly found in the Caribbean Sea and off the coast of Brazil.

LONG-SNOUTED SEAHORSE

ADULT HEIGHT
$3^{1}/_{2}$–7 inches (9–18 cm)

RINGS
11 + 37–39 (35–40)

PECTORAL FIN RAYS
16–18

DORSAL FIN RAYS
19–20 (17–20)

SPINES
medium to well-developed, blunt tips

CORONET
small but distinct, five rounded spines, long raised plate in front with spine at end

COLOR/PATTERN
dark green or brown, white spots coalescing into horizontal wavy lines

NOTABLE ANATOMY
adults usually with "mane" of thick skin filaments

DEPTH
shallow; overwinters in deeper water

HABITAT
seagrass beds, artificial structures

CONSERVATION STATUS
Data Deficient

Until recently the scientific name most often used for the Long-snouted Seahorse was *Hippocampus ramulosus*. However, after re-examination of the original type specimen of *H. ramulosus* in 1997, it was clear that it does not in fact match the Long-snouted Seahorse and was therefore being used in error. Baron von Cuvier was the first person to unambiguously name the Long-snouted Seahorse, and it is his species name (*H. guttulatus*) that we use today.

As its common name suggests, the Long-snouted Seahorse has a long snout (at least relative to the other European seahorse, *H. hippocampus*). It also usually sports a "mane" of thick skin fronds on its head and trunk. This develops only in mature animals (and may be lost later as the animal gets older), is found in both sexes and presumably helps camouflage them in their seagrass habitats. Another distinctive feature of this species are white spots on the body, which tend to coalesce into wavy white lines, often with dark edges. Individuals are generally dark green or brown in color.

H. guttulatus can be found in the waters around the UK (even as far north as Scotland), elsewhere in the Northeast Atlantic, and the Mediterranean through to the Black Sea. Genetically it is split into four subpopulations: Northeast Atlantic, southern Spain and Portugal, Mediterranean, and Black Sea. They prefer more thickly vegetated areas than the other eastern Atlantic/Mediterranean species, the Short-snouted Seahorse (*H. hippocampus*), and will use a variety of plants, sessile invertebrates, or artificial structures as holdfasts without an apparent preference.

The density of *H. guttulatus* is generally higher than *H. hippocampus*—in the Mar Piccolo di Taranto in southern Italy *H. guttulatus* were 14 times more abundant than *H. hippocampus*, in the Ria Formosa in Portugal they were ten times more abundant, and in dive surveys around Italy they were twice as abundant.

The breeding season is between March and October, gestation lasts for 21–28 days, and typical brood sizes range from 50–300 (maximum reported 581).

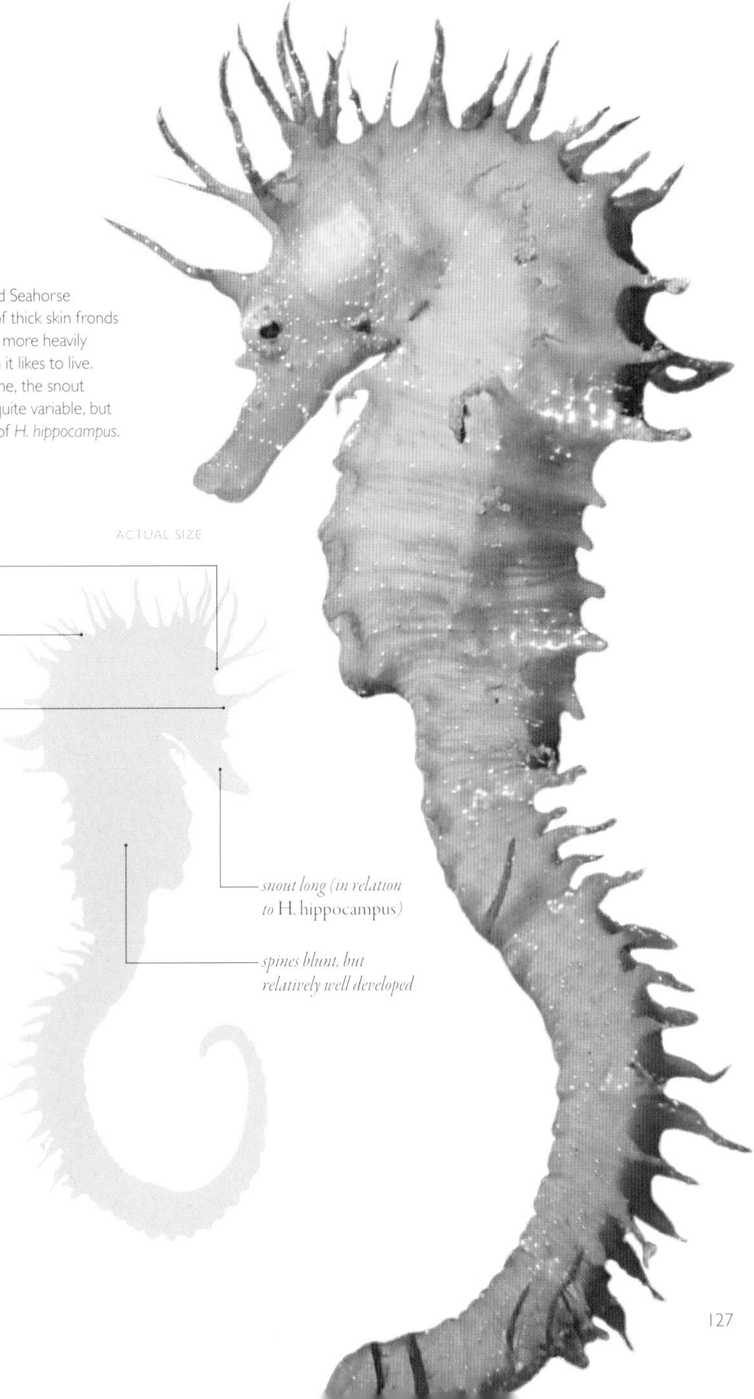

RIGHT The Long-Snouted Seahorse typically sports a mane of thick skin fronds that camouflage it in the more heavily vegetated areas in which it likes to live. Despite its common name, the snout length of this species is quite variable, but usually longer than that of *H. hippocampus*.

LINED SEAHORSE

Lined Seahorses occur in shallow waters off the east coast of North, Central and northern South America, from Nova Scotia to Brazil. They are quite variable in their morphology, especially their color, pattern, size, development of spines and coronet. As a result, they have been given many different names over the years. They can be identified, however, by the consistent (and unique) pattern of thicker trunk rings, and longer spines on trunk rings 1, 3 and 5, as well as by their fin ray, trunk and tail ring counts. Genetic evidence also points to them being a single, highly variable, species.

There is some question as to whether Lined Seahorses from the northern part of the range are resident, or migrants from further south. Genetic data, for those in New York at least, have confirmed that they are resident, although it is possible that they migrate offshore to deeper water during the colder months.

The Lined Seahorse is commonly bred as an aquarium fish, often under fancy names such as 'Sunburst,' 'Mustang,' 'Fire Red' and 'Sunfire.' Its breeding season in the wild is generally from May–October, and brood sizes usually range from about 200–500, although up to 1,300 young have been recorded.

ADULT HEIGHT
2½–7½ inches (6–19 cm)

RINGS
11 + 36 (34–39)

PECTORAL FIN RAYS
15–16 (14–18)

DORSAL FIN RAYS
18–19 (16–20)

SPINES
variable, none to well developed with blunt or sharp tips

CORONET
variable, low triangular wedge, ridge-like or raised with sharp edges or spines

COLOR/PATTERN
gray, brown, orange, yellow, red or black, distinctive horizontal lines on the trunk, and lines that follow the neck, often contrasting "saddles" across body

NOTABLE ANATOMY
trunk rings 1, 3, and 5 usually enlarged, with longer spines, snout usually < ½ head length

DEPTH
usually < 33 feet (10 m), but deeper in winter, max. depth recorded 240 feet (73 m)

HABITAT
seagrass, sponges, and floating *Sargassum*

CONSERVATION STATUS
Vulnerable

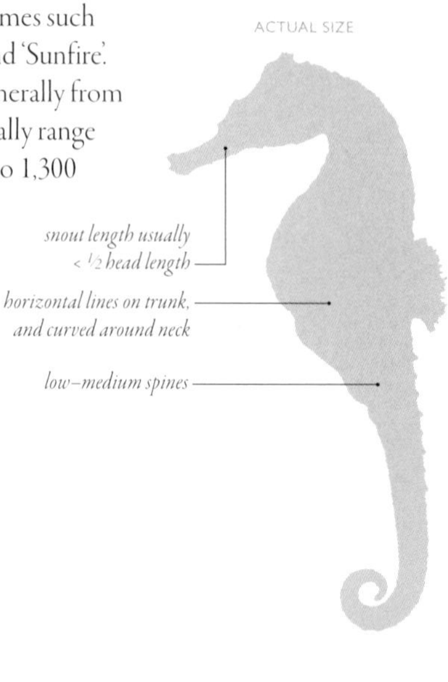

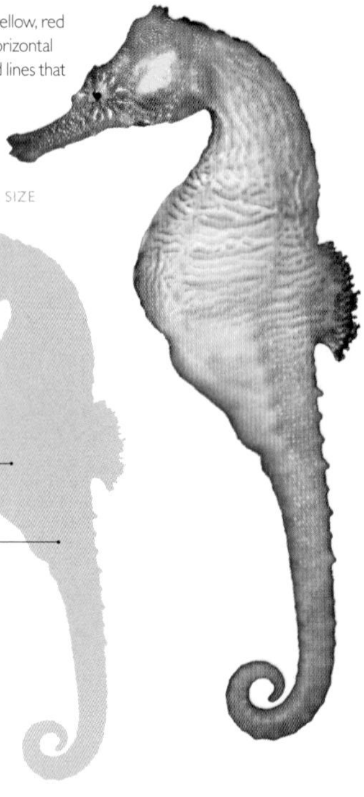

RIGHT The Lined Seahorse has distinctive horizontal lines on its trunk, and lines that follow the curve of the neck. These tend to be present regardless of the overall body color that may range from white to yellow, orange, red or black. Longer trunk spines on rings 1, 3 and 5 are also diagnostic.

Hippocampus hippocampus
(LINNEAUS, 1758)

SHORT-SNOUTED SEAHORSE

One of the most striking features of *H. hippocampus*, as its common name suggests, is its short snout, which is often less than a third of its head length. Preferring more open areas with sparse vegetation, the Short-snouted Seahorse actually does better in areas that have been trawled, and its populations tend to decrease as the habitat becomes re-vegetated. It is also often found in deeper areas that are open and exposed to oceanic currents.

Genetically, Short-snouted Seahorses exist as three distinct subpopulations: (i) English Channel and Bay of Biscay; (ii) Mediterranean, Iberian Peninsular and Macaronesian Islands; (iii) West Africa. Specimens from West Africa have much larger and more angular coronets, and may have diverged from their more northern cousins about 484,000 years ago. Some individuals of *H. hippocampus* have extensive branched skin filaments, especially on the head and trunk, and this often leads to their misidentification as *H. guttulatus*.

ADULT HEIGHT
$2\frac{3}{4}$–6 inches (7–15 cm)

RINGS
11 + 37 (35–38)

PECTORAL FIN RAYS
14 (13–15)

DORSAL FIN RAYS
17 (16–19)

SPINES
low

CORONET
narrow ridge or wedge-like, large and angular in West African specimens

COLOR/PATTERN
brown, yellow, orange, purple, or black with tiny white spots

NOTABLE ANATOMY
very short snout (usually $< \frac{1}{3}$ head length)

DEPTH
3–197 feet (1–60 m), overwinter in deeper water

HABITAT
soft-bottom, algal beds, rocky areas

CONSERVATION STATUS
Data Deficient

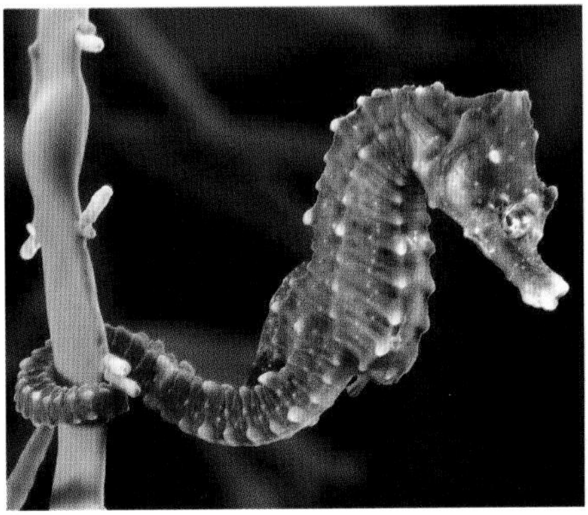

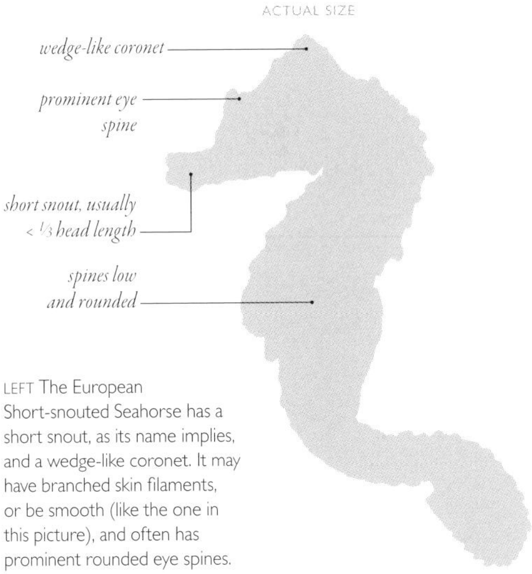

LEFT The European Short-snouted Seahorse has a short snout, as its name implies, and a wedge-like coronet. It may have branched skin filaments, or be smooth (like the one in this picture), and often has prominent rounded eye spines.

PATAGONIAN SEAHORSE

The Patagonian Seahorse was initially identified as a southern population of the Lined Seahorse. Upon further examination, however, Argentinian and southern Brazilian specimens turned out to be a separate species, that differs from *H. erectus* by 6 percent and from the eastern Atlantic *H. hippocampus* by 8 percent (mtDNA sequence).

The largest population of the Patagonian Seahorse is in San Antonio Bay, a semi-enclosed, estuarine, marine protected area in the northern part of the San Matías Gulf in Argentina. This species lives in shallow water, in association with red and green macroalgae, and is an opportunistic feeder. Its main foods are amphipods and crab larvae, however, juvenile seahorses have been found in the guts of several more mature individuals, showing that it is not averse to cannibalism.

ADULT HEIGHT
2½–6 inches (6–15 cm)

RINGS
11 + 37 (35–38)

PECTORAL FIN RAYS
14 (13–15)

DORSAL FIN RAYS
18 (16–18)

SPINES
low

CORONET
low ridge or wedge with sharp angles

COLOR/PATTERN
drab, pale to dark brown, or yellow, red, or orange, with irregular dark striations and small white dots on head and body

NOTABLE ANATOMY
short snout, long tail

DEPTH
3–23 feet (1–7 m)

HABITAT
soft bottom, red and green algae, also bivalve beds

CONSERVATION STATUS
Data Deficient

LEFT Patagonian seahorses are smoother that their more northerly neighbors, with a low coronet, and lack the Lined Seahorse's distinctive lines.

ACTUAL SIZE

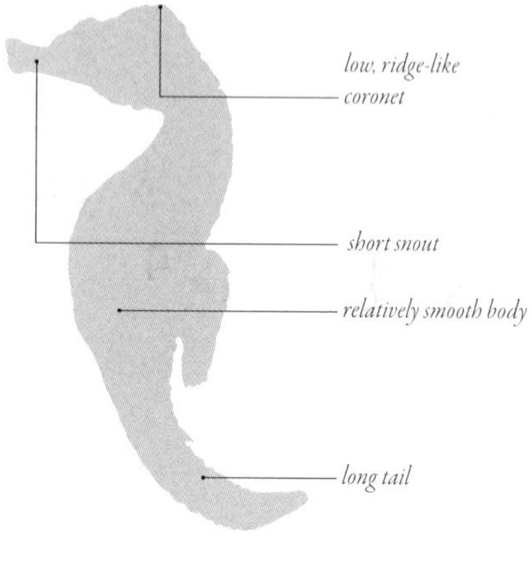

Hippocampus zosterae
JORDAN & GILBERT, 1882

DWARF SEAHORSE

The diminutive Dwarf Seahorse typically grows to less than 1¼ inches (3 cm) in height. It lives in shallow seagrass beds, and is relatively common on the Gulf coast of Florida. It can also be found in Texas, Mexico, and the Bahamas. In the wild it feeds primarily on tiny copepods and have been found to have as many as 200 in its digestive tract at one time.

Dwarf Seahorses mature rapidly, and are able to reproduce in less than three months from birth. Breeding takes place from February to October. Few live beyond one year, but some live for more than three years. Pair-bonding is usually between similar-sized individuals, and courtship involves an elaborate dance including quivering, reciprocal pointing, and rising before copulation takes place. The gestation period is about 12 days and a male will typically mate again within 4–48 hours of giving birth. Maximum brood size is about 50 young but usually broods are much smaller (<20).

ADULT HEIGHT
¾–1¼ inches (2–3 cm)

RINGS
9–10 + 31–32

PECTORAL FIN RAYS
11–12

DORSAL FIN RAYS
12

SPINES
low or knob-like

CORONET
high column or knob-like

COLOR/PATTERN
beige, yellow, green, black, with variable mottling

NOTABLE ANATOMY
short snout, usually < ⅓ head length, often covered in tiny warts

DEPTH
< 33 feet (10 m)

HABITAT
seagrass beds

CONSERVATION STATUS
Data Deficient

RIGHT The Dwarf Seahorse is very small (typically less than 1¼ inches (3 cm) in height) with a column-like coronet. Unlike the pygmy species, it is truly a miniature version of the larger seahorses, showing clear body rings and an external pouch in the males (as in the photograph).

ACTUAL SIZE

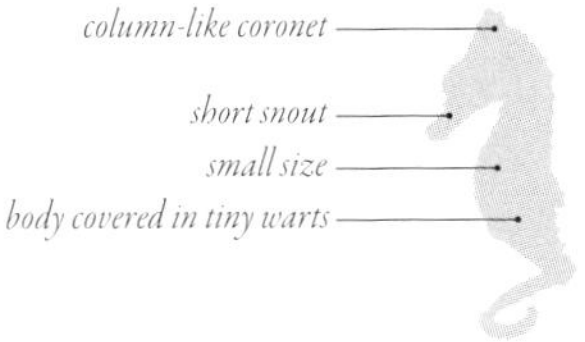

Hippocampus debelius

GOMON & KUITER, 2009

SOFT CORAL SEAHORSE

The Soft Coral Seahorse is a diminutive species, but may not be closely related to "true" pygmy seahorses. It has a distinct pouch on its tail, below the anal fin, separate gill openings, and a well-ossified skeleton. It has long spines on its head, trunk, and tail with dark bands around them. It is pale with numerous closely spaced vertical brown lines on its body, and has a distinctive coronet with three long spines. It was named after Helmut Debelius in recognition of his efforts to obtain specimens of this species.

It is found at depths of between 49–99 feet (15–30 m), in association with soft corals. These soft corals are also found in shallower water that is well shaded, so it is possible that this species may be found closer inshore.

ADULT HEIGHT
¾ inch (2 cm)

RINGS
10 + 28

PECTORAL FIN RAYS
10–11

DORSAL FIN RAYS
14

SPINES
long spines on several trunk and tail rings, above eye, and on coronet

CORONET
angular with three long spines

COLOR/PATTERN
pale background, closely spaced brown lengthwise stripes; dark band on spines

NOTABLE ANATOMY
separate gill openings, prominent long, slender spines

DEPTH
49–99 feet (15–30 m)

HABITAT
soft coral

CONSERVATION STATUS
Not Evaluated

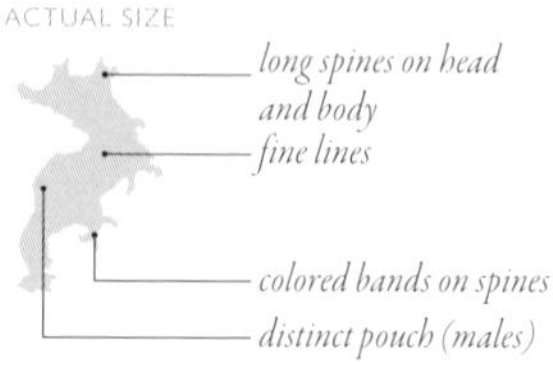

LEFT One of the most recently described seahorse species, the Soft Coral Seahorse has a striking color pattern, with distinctive fine lines and very long spines on its coronet, above the eyes, and on some trunk and tail rings.

Hippocampus jugumus
KUITER, 2001

COLLARED SEAHORSE

The Collared Seahorse is known only from three specimens: the holotype, which was found washed up on a beach on Lord Howe Island in 1925; a live specimen photographed at 66 feet (20 m) depth at Poor Knights Islands Marine Reserve in New Zealand in 2006; and most recently a specimen that came from the stomach of a captured Kingfish near Lord Howe Island in 2013.

It looks superficially similar to Fisher's Seahorse (*H. fisheri*), but has more trunk rings and a broad neck ring, which is confluent over its back, supporting a single mid-dorsal gill opening. Like Fisher's Seahorse, it has sharp, double eye spines, and is relatively small overall. It also looks like another poorly known species, *H. pusillus* (Dwarf Thorny Seahorse), but has fewer trunk rings, more tails rings, and more fin rays. Due to the limited interaction with this species, photographs are extremely rare.

ADULT HEIGHT
1½–2 inches (4–5 cm)

RINGS
12 + 38

PECTORAL FIN RAYS
16

DORSAL FIN RAYS
21

SPINES
medium, sharp; eye spine double; longer on TrR 1,4,7,11 and several TaR

CORONET
slightly raised with seven tiny points and medium-long spine in front

COLOR/PATTERN
pale orange-pink

NOTABLE ANATOMY
12 trunk rings; single gill opening

DEPTH
66 feet (20 m)

HABITAT
rocky reef

CONSERVATION STATUS
Not Evaluated

RIGHT Rarely encountered, the Collared Seahorse is relatively small, and slightly spiny, but with an extra trunk ring compared to most species. Very little is known of its habitat or life.

ACTUAL SIZE

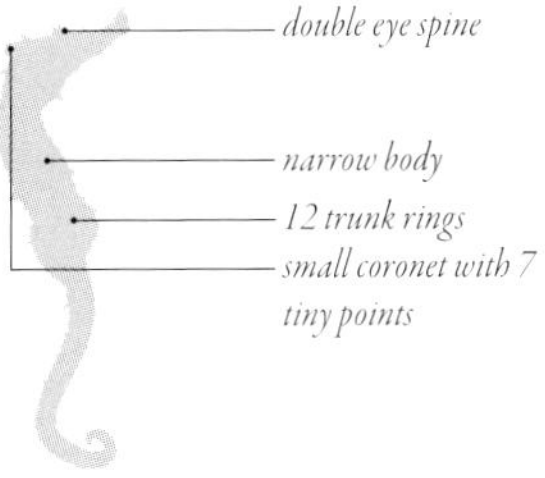

PYGMY THORNY SEAHORSE

The Pygmy Thorny Seahorse looks very similar to *H. jugumus* (Collared Seahorse), but it differs primarily in number of dorsal and pectoral fin rays. It is known only from three specimens from New Caledonia, all of which were collected at great depths (115–748 feet/ 35–228 m). They were dredged by French research vessels, R/V *Alis* (LIFOU, 2000) and R/V *Vauban* (MUSORSTOM 4, 1985), from areas of coral gravel, coralline rocks, and coral.

The specimens apparently (I have not examined them) have 13 trunk rings, a spiny exterior, and a raised coronet with six groups of spines, each group with one to six spines. The shoulder ring is confluent over the back of the neck, as in *H. jugumus*, however the gill openings are not fused. Nothing is known of this seahorse in life. This species has never been photographed alive.

ADULT HEIGHT
1¼–1½ inches (3–4 cm)

RINGS
13 + 34

PECTORAL FIN RAYS
12–13

DORSAL FIN RAYS
15

SPINES
moderately well-developed; double spine above eye

CORONET
slightly raised with six projections, each of which have one to six spines

COLOR/PATTERN
uniform yellow

NOTABLE ANATOMY
gill openings separate

DEPTH
115–748 feet (35–228 m)

HABITAT
coral

CONSERVATION STATUS
Not Evaluated

ACTUAL SIZE

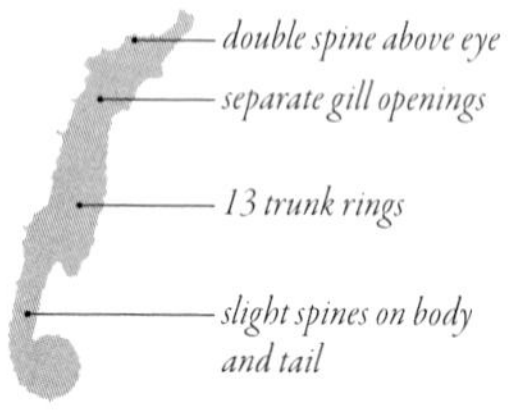

RIGHT The Pygmy Thorny Seahorse is a deep water species with 13 trunk rings and a spiny exterior. Nothing is known of its life history or ecology.

TYRO'S SEAHORSE

Tyro's Seahorse is named after the Dutch research vessel on which Jack Randall was participating in a biological expedition to the Seychelles in 1992 when the new seahorse was found. It is known from only one specimen, dredged from 141–158 feet (43–48 m) among coral fragments of *Stylophora pistillata*, *Montipora digitata* and *Dendrophyllia* sp. Unusually for a seahorse, it has 14 trunk rings. It also has a heavyset head, with a rounded hump behind the coronet. When first collected, it was light brown, with scattered white patches all over, and darker patches over the more enlarged dorsal spines of the fourth and seventh trunk rings. Leaf-like branching filaments were attached to larger dorsal spines of trunk and tail. No further specimens have been found, and nothing is known of its ecology or life history although it may be closely related to *H. jugumus* and *H. pusillus*, and possibly *H. fisheri*, and shares with *H. jugumus* the continuous shoulder ring and the single dorsal gill opening on the middle of the neck.

ADULT HEIGHT
2½ inches (6 cm)

RINGS
14 + 38

PECTORAL FIN RAYS
14–15

DORSAL FIN RAYS
15

SPINES
moderate and blunt

CORONET
sharply raised with large hump behind

COLOR/PATTERN
pale brown with white blotches, and darker blotches on dorsal surface of TrR 4, 7; more orange toward the tail tip

NOTABLE ANATOMY
heavyset head with hump behind the coronet, double nose spine, double eye spines, continuous shoulder ridge, single gill opening

DEPTH
141–157 feet (43–48 m)

HABITAT
coral

CONSERVATION STATUS
Not Evaluated

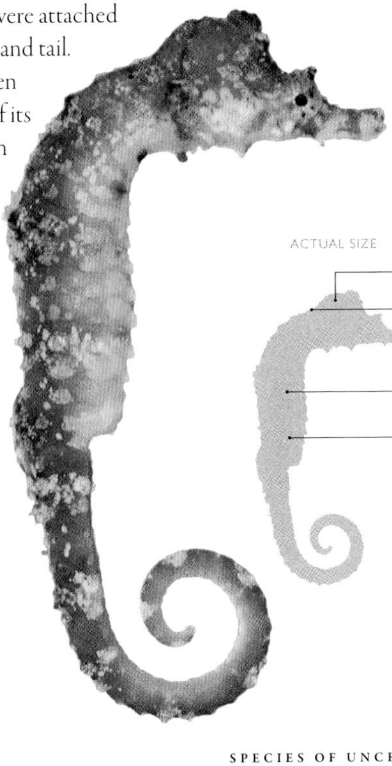

RIGHT Tyro's Seahorse has a heavyset head, the neck region of which is raised behind the coronet. It has more trunk rings than most seahorses and a head that seems to be less flexible than most species. Its blotched color pattern would camouflage it well against a sand or gravel seafloor.

ACTUAL SIZE

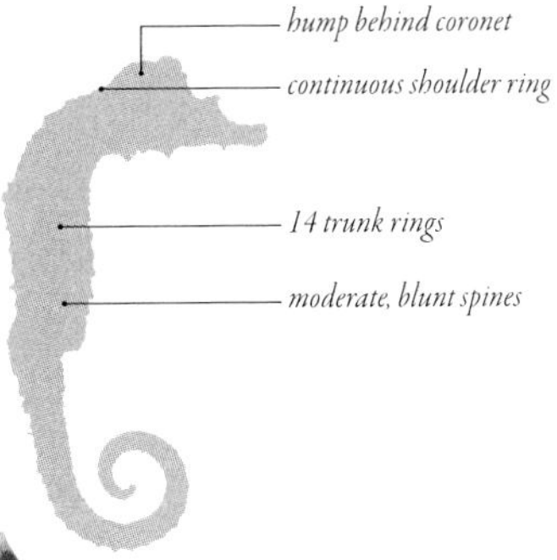

Hippocampus minotaur

GOMON, 1997

BULLNECK SEAHORSE

The Bullneck Seahorse has a huge head, a distinctive thick neck, and a more graceful body, that lacks recognizable body segments. It is named after the mythical Greek Minotaur, who had the head of a bull and the body of a human. It is extremely flattened, and has a tiny dorsal fin with only seven rays. Bony plates are absent in the ventral portion of this species, with the exception of a crescent-like process that extends from the position of the lateral ridge of the last trunk rung and may support the "belly" (and in males the brood pouch), and separate this from a narrow strip in which all the abdominal organs are found. It is closely related to *H. paradoxus*.

The three known specimens were trawled from 210–361 feet (64–110 m). Little is known of its mid-continental shelf habitat, although the paratypes (some of the specimens that were used in the original species description) were collected with "fine sand." This species has never been photographed alive.

ADULT HEIGHT
2 inches (5 cm)

RINGS
8 + 41

PECTORAL FIN RAYS
11

DORSAL FIN RAYS
7

SPINES
none, body totally flat

CORONET
low mound

COLOR/PATTERN
mostly cream or brownish, speckled with brown, cream-centered spots (preserved specimens)

NOTABLE ANATOMY
huge head, thick neck, fleshy body; no distinct rings, laterally flattened, very short snout

DEPTH
210–361 feet (64–110 m)

HABITAT
fine sandy or hard bottom

CONSERVATION STATUS
Data Deficient

ACTUAL SIZE

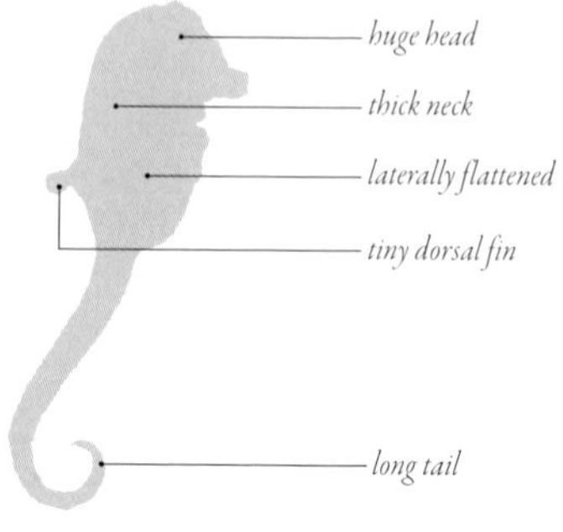

RIGHT The Bullneck Seahorse is known from only three specimens from southeast Australia. They were trawled from mid-continental shelf waters and nothing further is known of this species in life.

Hippocampus paradoxus

FOSTER & GOMON, 2010

PARADOXICAL SEAHORSE

The most recently described species is the Paradoxical Seahorse. Paradoxically the type specimen lay undisturbed in the South Australian Museum for 15 years prior to its publication in 2010.

H. paradoxus was dredged from 335 feet (102 m) during a geo-scientific expedition off the southwest coast of Australia, 1,243 miles (2,000 km) west of the known distribution of *H. minotaur*, by the CSIRO research vessel R/V *Franklin* at the far western end of the Great Australian Bight.

The fleshy nature of the specimen necessitated some ingenuity in order to examine it, and it is the first seahorse to have undergone CT scanning. It is also the only species entirely to lack a dorsal fin (assuming that this was not the result of damage during collection), which is extraordinary given that it is the dorsal fin that usually provides propulsion enabling seahorses to move. Its bone structure is much reduced with only eight body segments, a bulbous head, and is flattened laterally. It also has an series of fleshy, fin-like lobes along the mid-line of the trunk and tail. A live specimen has never been photographed.

ADULT HEIGHT
2½ inches (6 cm)

RINGS
8 + 41

PECTORAL FIN RAYS
11

DORSAL FIN RAYS
0

SPINES
none—small fleshy papillae in place of some spines on head

CORONET
not visible, mound-like head

COLOR/PATTERN
unknown in life, pale cream with dark dots in preservative

NOTABLE ANATOMY
very compressed, long tail, mound-like head, no dorsal fin, fleshy lobes on back of trunk and tail

DEPTH
335 feet (102 m)

HABITAT
sand, bryozoans, lace corals, sponges

CONSERVATION STATUS
Not Evaluated

ACTUAL SIZE

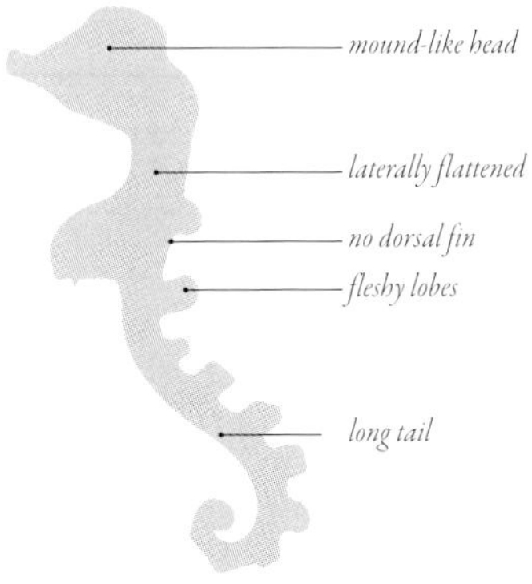

RIGHT The Paradoxical Seahorse appears to be closely related to the Bullneck Seahorse with which it shares many unusual features. Also obtained from great depth, we have no further information about its life.

ZEBRA SEAHORSE

The Zebra Seahorse has an extremely striking black and white striped pattern, and it can get confused with "zebra-striped" forms of the Three-spot Seahorse (*H. trimaculatus*). They can be told apart by the shape of their coronet—raised and pointed in *H. zebra* and almost non-existent in *H. trimaculatus*. *H. zebra* also has a distinctive pointed eye spine, whereas *H. trimaculatus* has sharp recurved eye and cheek spines. Tail ring and fin ray counts also differ between the species.

This seahorse is known from only a few specimens collected in deep water on the Great Barrier Reef and in Moreton Bay, Queensland, Australia, as well as specimens and observations that share morphology and color pattern that were found in Western Australia and were identified as *H. montebelloensis* (here treated as a synonym). There is no further information available about its life history or behavior.

ADULT HEIGHT
2¾–3¼ inches (7–8 cm)

RINGS
11 + 38–39

PECTORAL FIN RAYS
15–16

DORSAL FIN RAYS
17

SPINES
very low, small, and sharp, or none

CORONET
medium, conical, five to six tiny points

COLOR/PATTERN
striking black (or dark brown) and white stripes all over

NOTABLE ANATOMY
prominent sharp eye spine

DEPTH
66–230 feet (20–70 m)

HABITAT
coral reef

CONSERVATION STATUS
Data Deficient

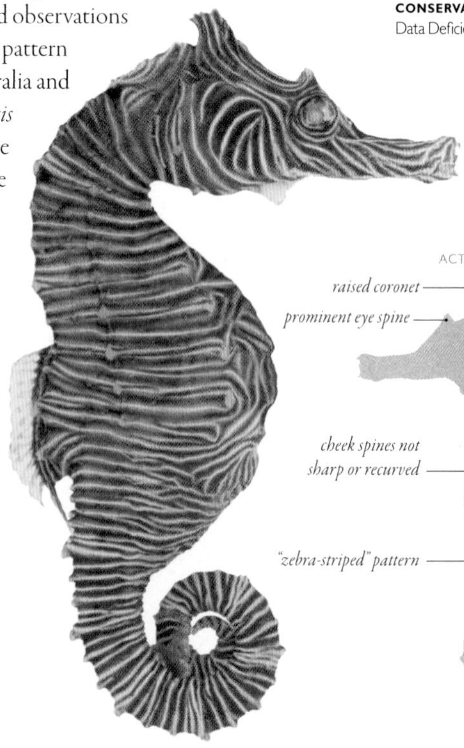

RIGHT The striking pattern of the Zebra Seahorse makes it one of the most distinctive seahorse species. It can however be confused with the "zebra-striped" form of the Three-spot Seahorse.

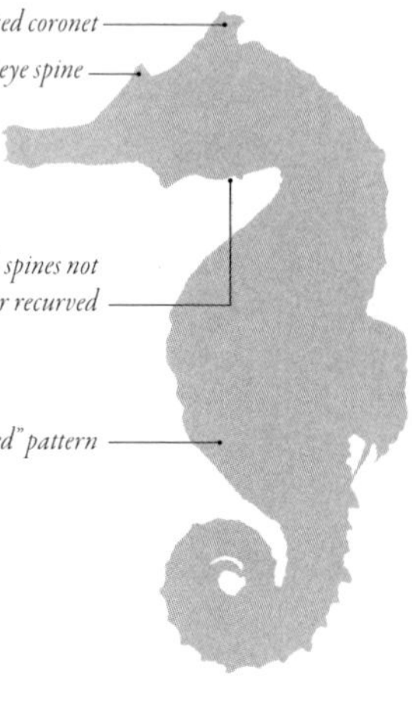

SEAHORSE RELATIVES

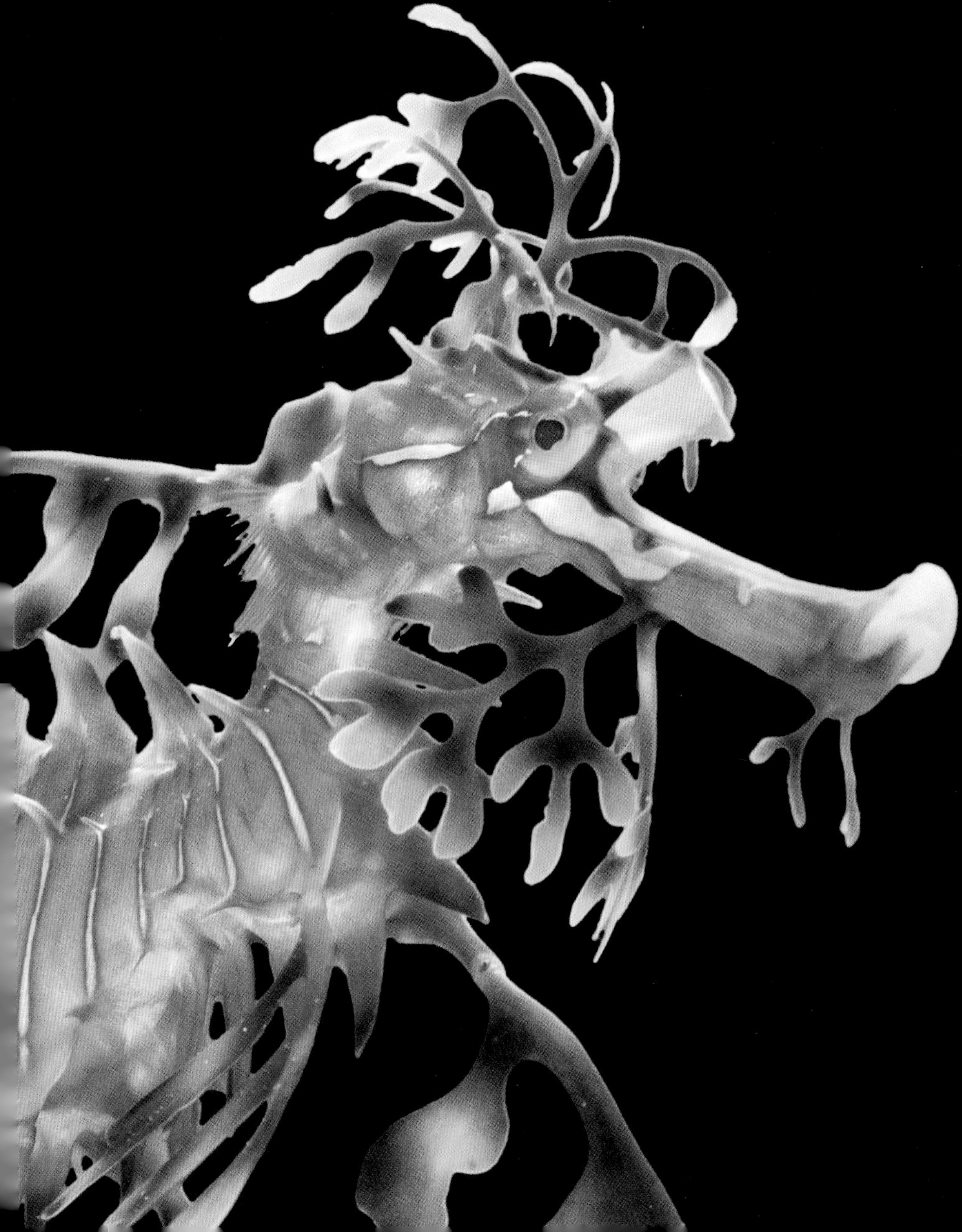

Nerophis ophidion
(LINNAEUS, 1758)

STRAIGHTNOSE PIPEFISH

The Straightnose Pipefish occurs in the eastern Atlantic, the Baltic, the Mediterranean, and the Black Sea. It is a temperate estuarine trunk-brooding (Gastrophori) species that is found as far north as Norway. It lives in seagrass beds, but is only resident there in the warmer months. In the winter it migrates to deeper water.

Fertilization of the eggs takes place within the female's body and then she deposits them in unprotected rows on the underside of the male's trunk where they are attached, aerated, and nourished. It is believed that this basic body plan, with no pouch plates to protect the developing young, is similar to that of an "ancestral" pipefish.

Each male receives eggs from a single female. A female, however, can transfer eggs to several males within the time that it takes for a male to brood the young. This type of mating system is called polyandry and is common in pipefish.

ADULT LENGTH
6–8 inches (15–20cm)

RINGS
28–33 + 68–74

PECTORAL FIN RAYS
0

DORSAL FIN RAYS
24–29

COLOR/PATTERN
green, bright blue stripes in females

NOTABLE ANATOMY
smooth skin; skin folds in breeding females

DEPTH
< 82 feet (25 m)

HABITAT
estuaries, algae or seagrass; sometimes enters freshwater

CONSERVATION STATUS
Least Concern

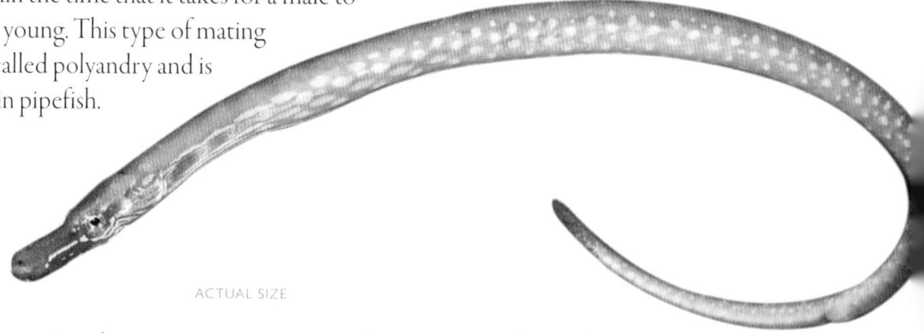

ACTUAL SIZE

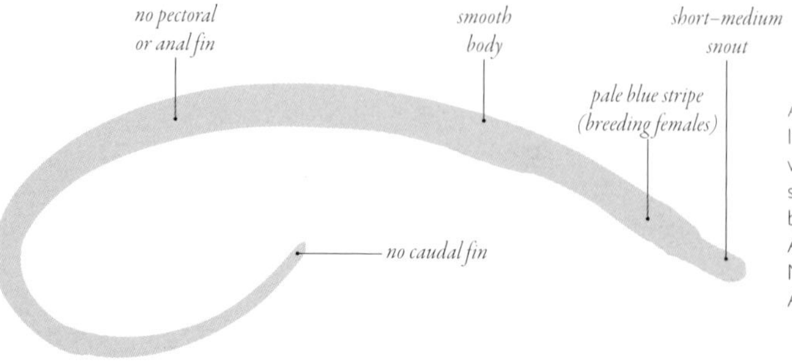

ABOVE Straightnose Pipefish live among algae and seagrass where their green color and semi-prehensile tail allow them to blend in easily. They breed from April–October and live in the Mediterranean, eastern North Atlantic, Black and Baltic Seas.

Dunckerocampus dactyliophorus
(BLEEKER, 1853)

BANDED PIPEFISH

The Banded Pipefish is widely distributed throughout the tropical and subtropical Indo-Pacific, from the Red Sea to Japan, Samoa and Australia. It is one of the most commonly observed pipefish, and is easily recognized as a result of its striking color pattern, as well as its large, colorful tail (caudal) fin. It lives in sheltered spots, in small caves in rocky areas and on coral reefs, rarely at depths greater than 66 feet (20 m).

It is territorial and can be found in the same place over the course of many months. Pairs of mated pipefish are commonly found together, and groups of juveniles are sometimes seen. The main threats to its populations are habitat loss and collection for the aquarium trade.

As with the Straightnose Pipefish (*Nerophis ophidion*), the eggs of the Banded Pipefish are glued to the trunk (Gastrophori) of the male for brooding, and are very obvious as several rows of bright pink spheres. They are not protected by skin folds or pouch plates.

ADULT LENGTH
3½–7 inches (9–18 cm)

RINGS
16 (15–17) + 18–22

PECTORAL FIN RAYS
18–22

DORSAL FIN RAYS
20–26

COLOR/PATTERN
white with dark red or brown stripes

NOTABLE ANATOMY
banded pattern, large tail fin, two spines on principal ridges of each ring

DEPTH
17–184 feet (5–56 m), usually < 66 feet (20 m)

HABITAT
coral reef

CONSERVATION STATUS
Data Deficient

BELOW The color pattern of the Banded Pipefish easily identifies them and they are commonly spotted (often in pairs) on coral reefs and in small caves under overhangs. Some disagreement exists as to the genus name of this species and sometimes it is classified as *Doryrhamphus dactyliophorus*.

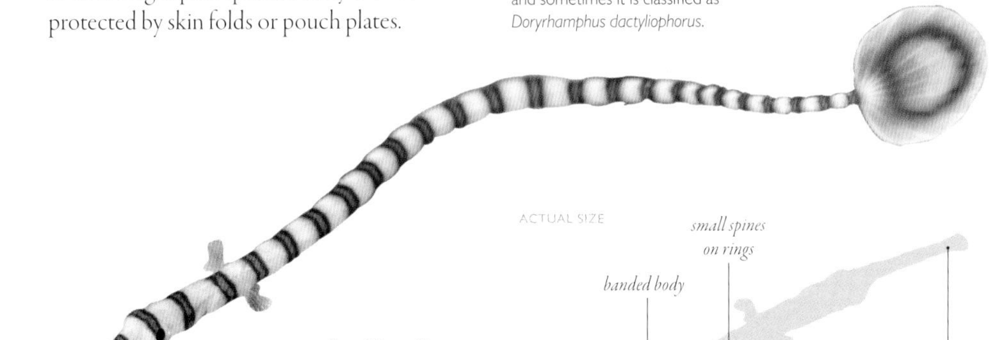

ACTUAL SIZE

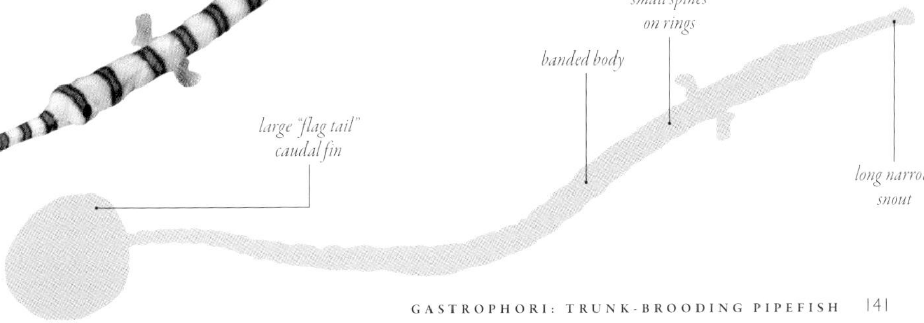

SCRIBBLED PIPEFISH

The Scribbled Pipefish is one of at least 10–23 different species of the Indo-Pacific genus of tail-brooding (Urophori) pipefish *Corythoichthys*. Numerically, *Corythoichthys* species make up the bulk of tropical Indo-Pacific pipefishes found in museum collections.

The Scribbled Pipefish lives in very shallow water (usually < 10 feet/3 m), and is often found on coral rubble, sand or in seagrass. It broods its eggs and young (usually 200–300) on its tail, and has folds of skin to help protect the developing young. However, these are not long enough to overlap one another, so the eggs in the middle of the brood remain in contact with the seawater.

Pairing with the same partner is known to last for at least a year and a half, but may be longer. In one study, a female who was 30 minutes late for the morning greeting left her partner visibly distressed. He rushed over to two other females who passed nearby, only to stop in his tracks about 6 inches (15 cm) from them, presumably realizing his mistake.

ADULT LENGTH
2¼–6¼ inches (6–16 cm)

RINGS
16 (15–17) + 33–35 (31–37)

PECTORAL FIN RAYS
16 (14–18)

DORSAL FIN RAYS
28–29 (26–32)

COLOR/PATTERN
pale with variable dark markings, 20 or more dark bands (made up of multiple fine lines) across body and tail

NOTABLE ANATOMY
narrow snout and sharply raised forehead

DEPTH
< 10 feet (3 m)

HABITAT
sand, coral, seagrass

CONSERVATION STATUS
Not Evaluated

BELOW The Scribbled Pipefish lives in shallow water and is typically found in sand, seagrass, or coral habitats. Mated pairs exhibit complex greeting and mating rituals and the male broods the eggs on a patch on his abdomen that is protected by flaps of skin.

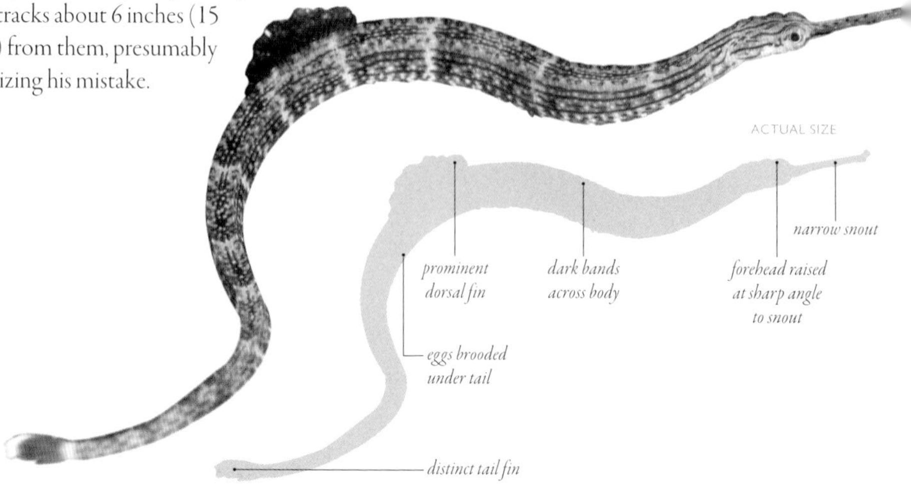

Halicampus macrorhynchus
(BAMBER, 1915)

ORNATE PIPEFISH

The Ornate Pipefish is one of about 12 species in the genus *Halicampus*. It is found throughout the Indo-Pacific, from east Africa and the Red Sea to the Solomon Islands. It has a series of about eight or nine pairs of large leaf-like skin flaps attached to the dorsal ridge of the body, a relatively long snout, and broods its eggs and young on the underside of its tail.

Specimens are generally found on sandy bottoms and coral rubble. Little is known of its ecology, life history, or behavior.

ADULT LENGTH
5¼–7 inches (13–18 cm)

RINGS
14 (14–15) + 25–27

PECTORAL FIN RAYS
17–19 (16–19)

DORSAL FIN RAYS
18–19

COLOR/PATTERN
variable, reddish-brown, greenish-yellow, c. ten dark bars crossing body

NOTABLE ANATOMY
skin flaps at regular intervals on body

DEPTH
10–82 feet (3–25 m), apparently deepest from trawl at 591–984 feet (180–300 m)

HABITAT
coral rubble, algae-covered rocks

CONSERVATION STATUS
Not Evaluated

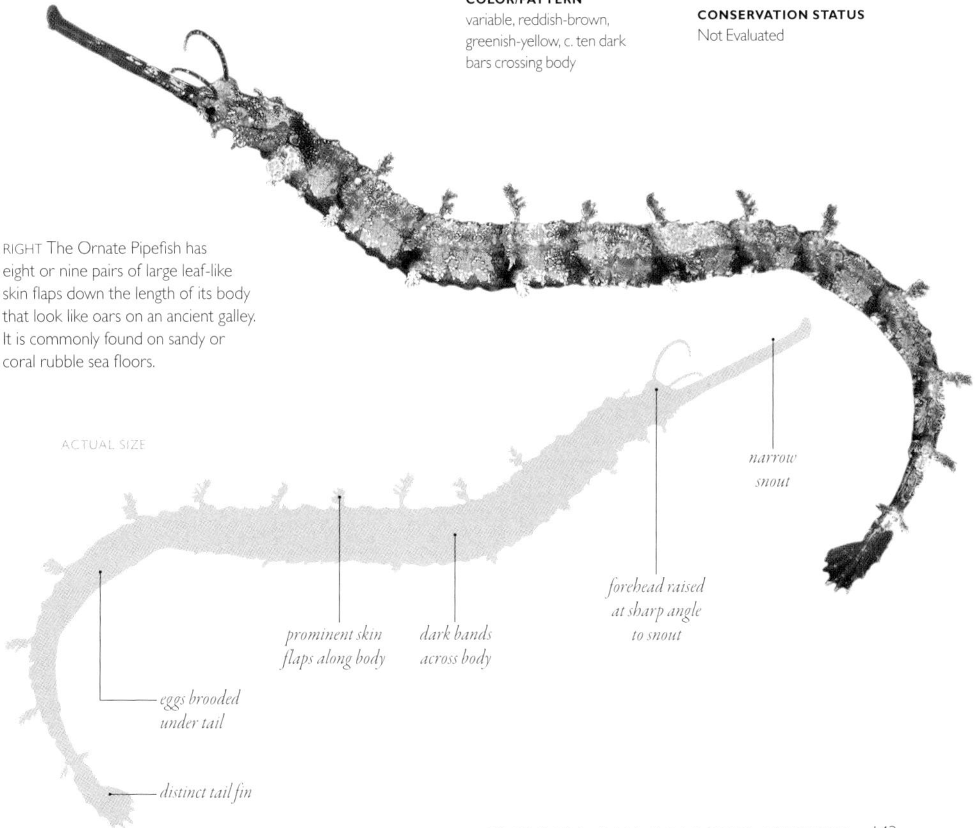

RIGHT The Ornate Pipefish has eight or nine pairs of large leaf-like skin flaps down the length of its body that look like oars on an ancient galley. It is commonly found on sandy or coral rubble sea floors.

Haliichthys taeniophorus
GRAY, 1859

RIBBONED SEADRAGON

Known as the Ribboned Seadragon, Ribboned Pipehorse, or Ribboned Pipefish, *Haliichthys taeniophorus* is not in fact closely related to either the seadragons or to the pipehorses. Instead, it is a Urophori (tail-brooding) pipefish, which is related to the Bent Stick Pipefish (*Trachyrhamphus bicoarctatus*), the pygmy pipehorses (for example *Acentronura*), and the seahorses (*Hippocampus*). It is found only in northern and western Australia.

In life, this species has extensive skin filaments, which are reminiscent of true seadragons. Preserved specimens tend to lose these branching filaments, but they retain their spiny exterior, whereby the junction of each plate of body armor is extended into a sharp, curved spine. Juveniles have a tail (caudal) fin, but this is lost in adults.

ADULT LENGTH
6¼–11¾ inches (16–30 cm)

RINGS
18–20 + 43–47

PECTORAL FIN RAYS
20–21 (19–22)

DORSAL FIN RAYS
24–27

COLOR/PATTERN
usually greenish-yellow with variable blotches; deeper water specimens tend to be more reddish-brown

NOTABLE ANATOMY
long snout, spiny body and tail, branched skin filaments

DEPTH
< 66 feet (20 m)

HABITAT
weedy areas bordering open substrates

CONSERVATION STATUS
Not Evaluated

BELOW The Ribboned Pipefish looks superficially like a seadragon with its leafy appendages, however it is more closely related to plainer pipefish. Its skin fronds help to camouflage it in its weedy habitats.

ACTUAL SIZE

Siokunichthys nigrolineatus
DAWSON, 1983

MUSHROOM CORAL PIPEFISH

This small pipefish species lives among the long tentacles (polyps) of mushroom corals, hence its common name, the Mushroom Coral Pipefish. It is distributed in the Coral Triangle region of the Philippines, Indonesia, and Papua New Guinea. Usually it can be found alone or in small groups on solitary corals, often at the bottom of shallow drop-offs, and frequently on reefs near large lagoons. Groups of Mushroom Coral Pipefish generally comprise a mix of individuals of varying ages and sex, but nothing else is known about their ecology or biology. They are very active and constantly move among the tentacles of their host mushroom coral.

This species is not collected for medicinal purposes, nor as an aquarium fish, and its fungiid coral hosts are considered to be among the least threatened hard corals. As a result it is listed as Least Concern in the IUCN Redlist.

ADULT LENGTH
2¼–3½ inches (6–9 cm)

RINGS
14 + 53–55

PECTORAL FIN RAYS
10–12

DORSAL FIN RAYS
30–32

CAUDAL FIN RAYS
10

COLOR/PATTERN
plain white with stripe through eye

NOTABLE ANATOMY
red, brown, or black diagonal stripe through eye

DEPTH
43–66 feet (13–20 m)

HABITAT
mushroom corals

CONSERVATION STATUS
Least Concern

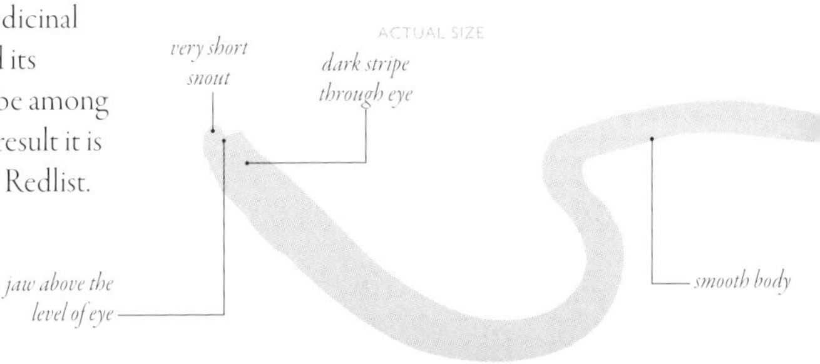

RIGHT The Mushroom Coral Pipefish lives permanently among the tentacles (polyps) of mushroom corals (especially *Fungia* spp.). It is very active and is usually seen swimming in and out of the tentacles.

Stipecampus cristatus
(MCCULLOCH & WAITE, 1918)

RINGBACK PIPEFISH

The Ringback Pipefish is the only species in the genus *Stipecampus*, and it is endemic to (found only in) southeast Australia. It occurs in South Australia, Victoria, and Tasmania among red and brown algae in sheltered reef habitats at depths of up to about 49 feet (15 m). It has a striking color pattern with pale blotches on a darker brown background, and a raised ridge on its snout that is confluent with the ridges over each eye. Its upper jaw is above the level of the eye resulting in a distinctly "grumpy" look. The males brood their young on their tail, where they are protected by pouch plates and skin folds. The original type specimen was taken by a dredge, but most specimens have been collected by SCUBA in 10–17 feet (3–5 m) of water.

ADULT LENGTH
7½–8½ inches (19–22 cm)

RINGS
19–21 + 39–42

PECTORAL FIN RAYS
12 (10–13)

DORSAL FIN RAYS
26–30

CAUDAL FIN RAYS
8

COLOR/PATTERN
brown to yellowish-brown with paler bars or blotches

NOTABLE ANATOMY
raised snout ridge, short snout

DEPTH
< 49 feet (15 m)

HABITAT
brown and red algae in sheltered reefs

CONSERVATION STATUS
Not Evaluated

BELOW Ringback Pipefish are dark brown with numerous pale patches over their back and sides. They are found only in southeast Australia, usually among red or brown algae. They are sometimes called Crested Pipefish on account of the ridge above their snout.

ACTUAL SIZE

GREAT PIPEFISH

The Great Pipefish is found from northwest Africa to the Baltic and Black Seas. Records from South Africa may represent a different species. Described by Linneaus in his *Systema Naturae*, this is one of the first pipefish to have been named. It is a relatively large, temperate, marine species, which sometimes enters brackish water in estuaries. Populations in different areas have very different maximum sizes (17¼ inches/45 cm in Sweden versus 10¼ inches/26 cm in Turkey), and this also has a corresponding effect on fecundity and brood sizes.

The Great Pipefish has a completely enclosed pouch, and it used to be thought that it was closely related to seahorses, but the mechanism of closure is different. In *Syngnanthus* species, folds of skin overlap and are closed by a "zipper-like" mechanism down the front.

Maximum brood size is about 400 and gestation lasts for four to five weeks. Egg size is positively correlated to female body size (large females produce larger eggs), and eggs are significantly larger in this species than in species that lack brood pouches (for example, *Nerophis ophidion*).

ADULT LENGTH
2¼–17¾ inches, 2¼–10¼ inches in Turkey (12–45 cm, 6–26 cm in Turkey)

RINGS
19 (18–21) + 36–43

PECTORAL FIN RAYS
12–13 (10–14)

DORSAL FIN RAYS
33–42

COLOR/PATTERN
light greenish to brownish-gray, with paler or darker bands across body

NOTABLE ANATOMY
large body, long snout

DEPTH
usually < 49 feet (15 m), but recorded from up to 295 feet (90 m)

HABITAT
seagrass, algae

CONSERVATION STATUS
Not Evaluated

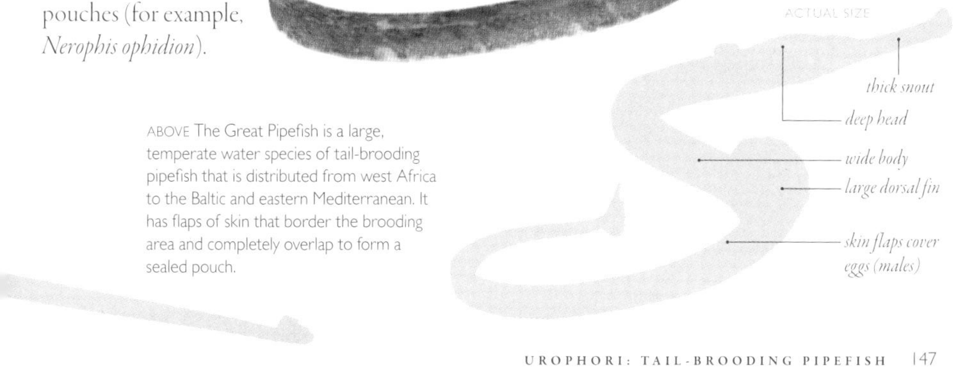

ABOVE The Great Pipefish is a large, temperate water species of tail-brooding pipefish that is distributed from west Africa to the Baltic and eastern Mediterranean. It has flaps of skin that border the brooding area and completely overlap to form a sealed pouch.

BENT STICK PIPEFISH

The Bent Stick Pipefish, originally described from Indonesia by Pieter Bleeker, is long, stick-like, and lacks body appendages at maturity. It is closely related to the Ribboned Pipehorse (*Haliichthys taeniophorus*) and pygmy pipehorses, and this entire group is related to (sister to) seahorses. Its head is usually tipped at an angle to its body in fish longer than about 6 inches (15 cm), and it is commonly found on sandy or muddy bottoms throughout the Indo-Pacific, from east Africa and the Red Sea, to Japan, Australia, and New Caledonia. The tail (caudal) fin is rather long in juveniles, but may be lost, regenerated, or rudimentary in adults. Skin folds and pouch plates protect the developing eggs on the underside of the male's tail.

Planktonic juveniles (up to about 4 inches/10 cm) have six pairs of skin appendages attached to their dorsal surface, which may function as flotation devices, and are lost as they mature. Adults generally live in muddy or sandy habitats, up to about 131 feet (40 m) deep. They are collected incidentally by trawlers in India, and elsewhere, operating at depths c. 66 feet (20 m).

ADULT LENGTH
10¼–15¾ inches (26–40 cm)

RINGS
21–24 + 55–63

PECTORAL FIN RAYS
15–19

DORSAL FIN RAYS
24–32

COLOR/PATTERN
near white, yellow and orange, gray to almost black with lighter patches along body

NOTABLE ANATOMY
smooth body, head bent at slight angle in adults

DEPTH
3–138 feet (1–42 m)

HABITAT
sand, rubble, reef, seagrass

CONSERVATION STATUS
Not Evaluated

BELOW Bent Stick Pipefish live on sandy and muddy bottoms, sometimes among seagrass or on reefs. They have a bent body and "sit" on their tail to ambush prey. Males brood their eggs under their tails, protected by pouch plates and folds of skin.

ACTUAL SIZE

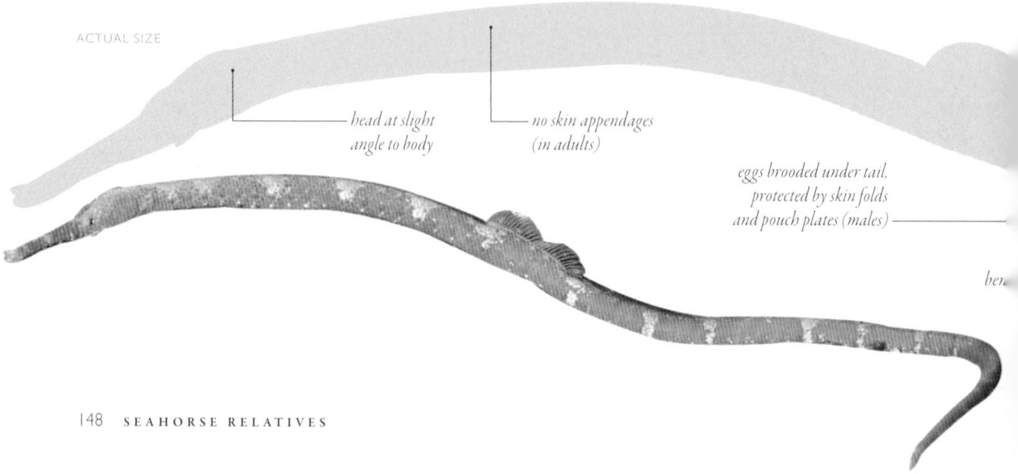

PORT PHILLIP PIPEFISH

The genus *Vanacampus* is endemic to (found only in) southern Australia, and comprises four species. The most widespread of these, the Port Phillip Pipefish, is found from the west to the east coasts in shallow "weedy" habitats or *Zostera* seagrass beds, often in estuaries and sheltered bays. It is a common species in, and was first described from, Port Phillip Bay near to Melbourne (hence its name), and is also common in rock pools and estuaries in Tasmania.

It is a tail-brooding pipefish that has up to about 50 eggs per brood. Brooding males occur during six to nine months of the year with the young joining the population in the Austral summer and early fall. It is a robust-looking pipefish, with a deep body and medium–long snout, and is reported to be a relatively strong swimmer and an active predator.

ADULT LENGT9H
3½–7½ inches (9–19 cm)

RINGS
18 (17–20) + 39–45 (38–46)

PECTORAL FIN RAYS
10–11 (9–12)

DORSAL FIN RAYS
22–29

CAUDAL FIN RAYS
10

COLOR/PATTERN
underside pale in males, and with dark streaks in females, dark margined pale blue bars on trunk and beginning of tail

NOTABLE ANATOMY
medium length snout, blue bars on side of body, underside of head pale

DEPTH
< 82 feet (25 m)

HABITAT
shallow sandy seagrass beds, estuaries, rock pools

CONSERVATION STATUS
Not Evaluated

ACTUAL SIZE

LEFT Port Phillip Pipefish are voracious predators and active swimmers. They tend to be darker on the dorsal and lighter on the ventral sides, and are often marked with pale blue on the sides.

Phycodurus eques

(GÜNTHER, 1865)

LEAFY SEADRAGON

The Leafy Seadragon, with its permanent leaf-like skin appendages and near perfect mimicry of drifting algae, is one of the most spectacular species of syngnathid. It is endemic to (found only in) southern Australia (from Wilson's Promontory in Victoria, to the Abrolhos Islands in Western Australia), and is typically found on rocky reefs, dominated by algae, at depths of 66–82 feet (20–25 m). South Australia has adopted the Leafy Seadragon as its state marine emblem, in recognition of its uniqueness and the fact that the largest populations of this species are found in South Australia.

The male broods eggs and developing young on a brood patch that is on the underside of the tail, and is open to the elements. A typical clutch may contain up to 250 eggs and be brooded for four weeks. Offspring are about ¾ inch (2 cm) long at birth, but reach maturity and adult size within 12–18 months.

The Leafy Seadragon grows to nearly 17 inches (44 cm), and it is highly sought after by public aquaria that can give it the space it needs. Owing to its protected status in Australia, only a single aquarium facility has a permit to collect a single pregnant male from the wild every year, in order to raise the young in captivity.

Acoustic tracking has been used to learn about the movements of these remarkable and threatened animals, and their home ranges are estimated at approximately 0.02 square miles (5 ha).

ADULT LENGTH
< 16¾ inches (43 cm)

RINGS
18 + 41–44

PECTORAL FIN RAYS
19–21

DORSAL FIN RAYS
34–38

COLOR/PATTERN
pale to brown, head with dark bars (allowing individual recognition), skin flaps shaded with brown broad lines across body

NOTABLE ANATOMY
long, leaf-like skin appendages, angular body, very long snout, no caudal fin

DEPTH
66–82 feet (20–25 m)

HABITAT
algae-covered rocky reefs

CONSERVATION STATUS
Near Threatened

LEFT A male Leafy Seadragon carrying eggs in the Gulf of St Vincent, South Australia. Its elaborate leafy skin appendages make this species look exactly like a piece of drifting seaweed. Larger than seahorses, but slightly smaller than the Weedy Seadragon, it has several long spear-like spines on either side of the body that serve as protection against predators.

Phyllopteryx taeniolatus
LACEPÈDE, 1804

WEEDY SEADRAGON

The Weedy Seadragon has a bent body shape like that of its close relative, the Leafy Seadragon (*Phycodurus eques*). It too is only found in Australia, from New South Wales to Western Australia, including Tasmania. It is more common than the Leafy Seadragon, but even in its favored rocky reef/algal habitat the Weedy Seadragon can only be found at densities of 10–70 individuals per 0.004 square miles (1 ha).

Once Weedy Seadragons settle down, at about four months of age and 6 inches (15 cm) in length, they will remain in the same location for the rest of their lives (up to ten years in Tasmania). They are usually solitary, however can be found in pairs and groups. They grow fast, reaching 11¾ inches (30 cm) in a year. At this stage, they are sufficiently differentiated to tell the sexes apart, although most individuals probably do not mate until their second year. The breeding season lasts from July–January. Pair-bonding has not been observed, and a female can develop a new batch of eggs in only two weeks and thus mate with another male before the first has given birth. Gestation usually lasts 30–38 days, and some males can breed twice in a season. Each brood contains approximately 100 young, about ¾ inch (2 cm) long at birth, with their distinctive skin appendages already developed.

ADULT LENGTH
11¾–17¾ inches (30–45 cm)

RINGS
17–18 + 31–37

PECTORAL FIN RAYS
21–22 (20–23)

DORSAL FIN RAYS
27–34

COLOR/PATTERN
brown, with white spots on head, trunk with dark bars and profusely spotted, dermal flaps pale with dark edges

NOTABLE ANATOMY
bent body, 12 long rounded dermal flaps, very long snout, no caudal fin

DEPTH
10–164 feet (3–50 m)

HABITAT
algal beds, rocky reefs

CONSERVATION STATUS
Near Threatened

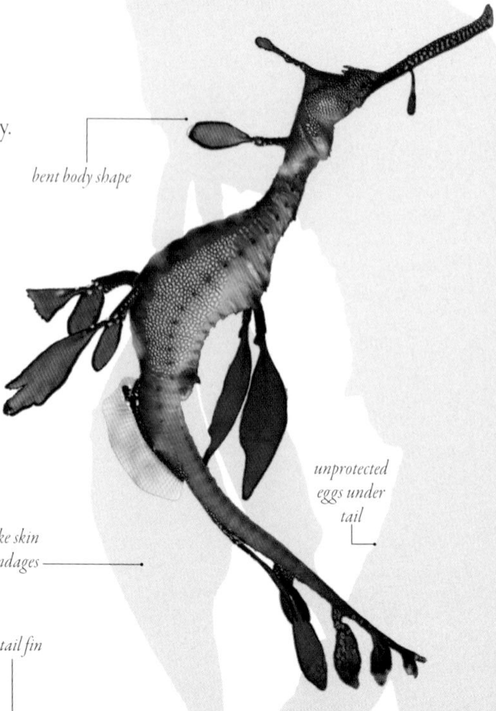

RIGHT The Weedy Seadragon has a distinctive bent body and fabulous markings on its trunk. Its leaf-like appendages help to camouflage it in its weedy habitat. Like the Leafy Seadragon it is found only in southern Australia.

Solegnathus spinosissimus
GÜNTHER, 1870

AUSTRALIAN SPINY PIPEHORSE

The Australian Spiny Pipehorse is restricted to the southeast part of Australia, from Tasmania to southern Queensland, and New Zealand. It is found in soft-bottom habitats, often close to rich invertebrate fishing grounds. As a result, it is frequently captured in trawls operating at depths of 99–820 feet (30–250 m). However, shallow-water populations (7–10 feet/2–3 m) occur in shaded waters near the Derwent Estuary in Tasmania. One collection was taken at 1,312–1,805 feet (400–550 m) depth.

The surface of the body is spinulose, meaning that it is covered in many tiny spines (*spinosissimus* means "very spiny"), and it has a prehensile, finless tail. Eggs are glued directly onto the underside of males' tails, and are not protected by brood plates or skin folds. Some individuals have been observed to breed when they grow to be 7¾ inches (20 cm), but most specimens with broods are > 11¾ inches (30 cm) in length. Typical brood sizes range from 20–200 young.

ADULT LENGTH
7¾–19½ inches (20–50 cm)

RINGS
25–26 (24–27) + 51–59

PECTORAL FIN RAYS
23–26

DORSAL FIN RAYS
34–42

COLOR/PATTERN
plain, with seven dark bars on back of trunk (may be reduced to paired dark spots)

NOTABLE ANATOMY
body surfaces spinulose (covered in tiny spines)

DEPTH
7–1,805 feet (2–550 m), mostly 99–820 feet (30–250 m)

HABITAT
rubble reefs, rocky reefs, sea whips, gorgonians, encrusting animals

CONSERVATION STATUS
Data Deficient

ABOVE The Australian Spiny Pipehorse looks like a naked seadragon. It has a similar bent shape, but instead of leafy skin appendages, it has spiny armor. It has a long snout and lives in areas of rubble and rocky reefs.

ACTUAL SIZE

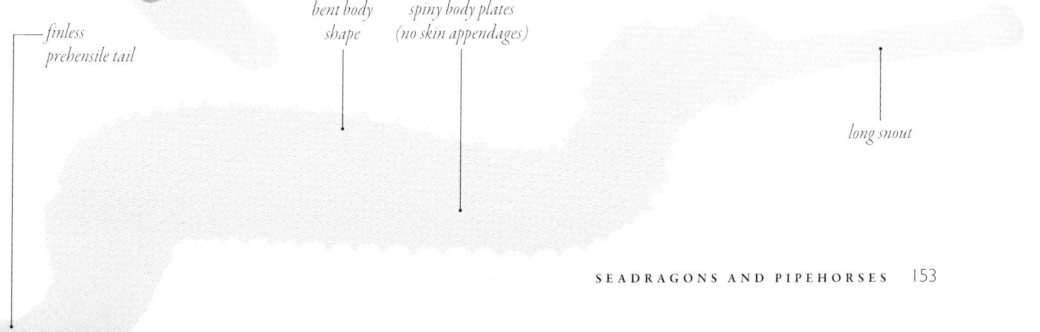

Syngnathoides biaculeatus

(BLOCH, 1785)

ALLIGATOR OR DOUBLE-ENDED PIPEFISH

The Alligator (or Double-ended) Pipefish is found in shallow seagrass beds, and is widely distributed from East Africa and the Red Sea, to southern Japan, Australia and Tonga. Its body is broadly triangular in cross section, and the end of its finless tail is prehensile. Interestingly, it appears that prehensile tails have evolved several times within the family Syngnathidae, and the Alligator Pipefish's prehensile tail is not from the same stock as seahorses.' The sexes are similar in size and shape (although males may be slightly longer), but can be told apart since females have white zig-zag lines on their abdomens, and blue spots, whereas males do not.

The eggs and developing embryos are brooded on the male's trunk, thus the alligator pipefish has been classified within the Gastrophori, however, genetic studies align it with the Urophori.

In the tropics the Alligator Pipefish breeds year-round. However, in subtropical Australia it breeds only in the summer months, from October to April. Females produce batches of 60–350 eggs that are transferred to a single male only (monogamy), and he broods them for 16–25 days, until the young are born.

ADULT LENGTH
7–11 inches (18–28 cm)

RINGS
16 (15–18) + 40–54

PECTORAL FIN RAYS
20–24

DORSAL FIN RAYS
38–48

COLOR/PATTERN
adults green with variable dark markings, females with white zig-zags on ventral surface and blue spots, juveniles brown

NOTABLE ANATOMY
broad ventral and narrow dorsal surfaces, so almost triangular in cross-section, prehensile tail tip, long snout

DEPTH
< 49 feet (15 m)

HABITAT
seagrass

CONSERVATION STATUS
Data Deficient

BELOW The Alligator Pipefish has a wide, flat underside and a back that is narrow, so in cross-section it looks like a trapezoid. It carries its young below the trunk (as in the picture), however this trait has been secondarily acquired as all its relatives are tail-brooders. The tip of its tail is prehensile.

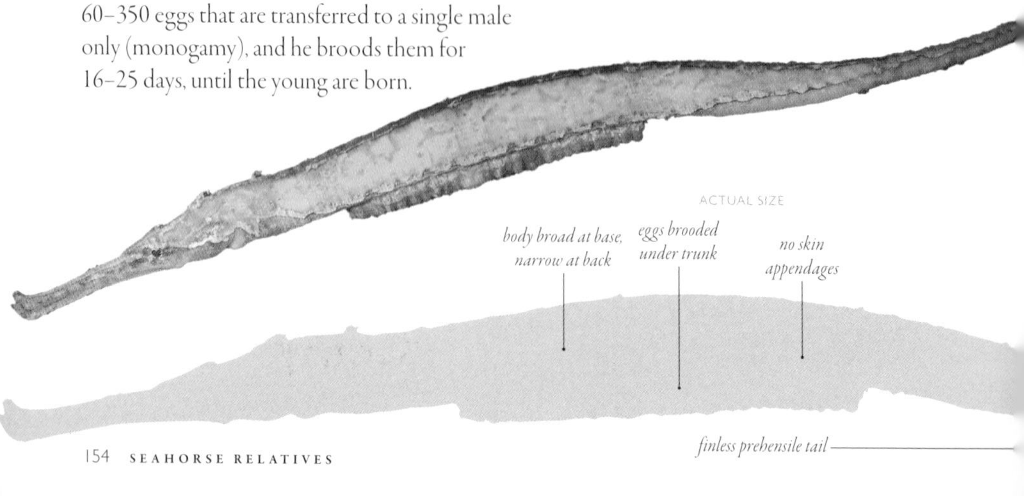

Acentronura breviperula

TEMMINCK & SCHLEGEL, 1850

SHORT-POUCH PYGMY PIPEHORSE

Morphologically in between a seahorse and a pipefish, the Short-pouch Pygmy Pipehorse is one of the species commonly known as a pygmy pipehorse. It is found in the central Indo-Pacific, and has an enclosed pouch and a prehensile tail like a seahorse, but a head that is held at less of an angle to its body, and has a poorly developed coronet.

Other pygmy pipehorse species include *A. tentaculata* in the Red Sea, *A. mossambica* in the western Indian Ocean, *A. gracilissima* in Japan and China, *Amphelikturus brachyrhynchus* in the Atlantic, as well as three species in Australia—*Idiotropiscis larsonae*, *I. australe*, and *I. lumnitzeri*. Originally part of the genus *Acentronura*, specimens referred to as *Idiotropiscis* and *Kyonemichthys* are now treated as separate genera. All species live in shallow water, and are rarely seen. They are primarily known from a few specimens in museum collections, and from keen-eyed divers. Further research is needed to understand more about these elusive creatures.

ADULT LENGTH
$2\frac{1}{2}$–4 inches (6–10 cm)

RINGS
13 + 44–46

PECTORAL FIN RAYS
14–17

DORSAL FIN RAYS
16

COLOR/PATTERN
greenish brownish-red

NOTABLE ANATOMY
head held at slight angle to body, prehensile tail

DEPTH
< 131 feet (40 m)

HABITAT
red algae, rocky reefs

CONSERVATION STATUS
Not Evaluated

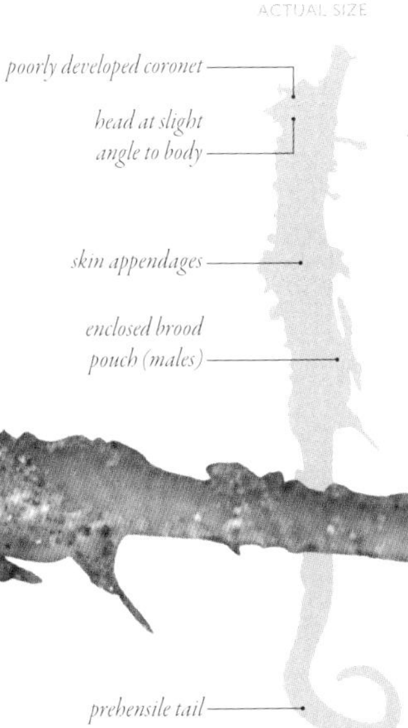

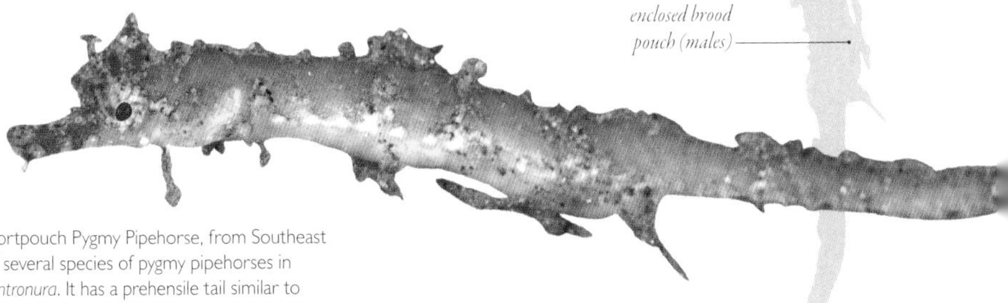

RIGHT The Shortpouch Pygmy Pipehorse, from Southeast Asia, is one of several species of pygmy pipehorses in the genus *Acentronura*. It has a prehensile tail similar to seahorses, with which it grabs a holdfast; however, unlike in seahorses, its body is more straightened. This one is from northern Sulawesi, Indonesia.

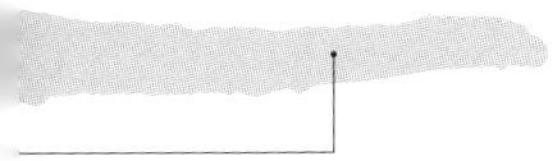

REFERENCES AND FURTHER READING

REFERENCES

DAWSON, C.E. (1985) *Indo-Pacific Pipefishes*. Gulf Coast Research Laboratory, Ocean Springs, Mississippi, USA.

FOSTER, S.J. & VINCENT, A.C.J. (2004) *Life history and ecology of seahorses: implications for conservation and management*. Journal of Fish Biology, 64(6): 1–61.

KUITER, R.H. (2009) *Seahorses, Pipefishes and their Relatives*. Aquapress, Seaford, Australia.

LOURIE, S.A, VINCENT, A.C.J. & HALL, H.J. (1999) *Seahorses: an identification guide to the world's species and their conservation*. Project Seahorse.

LOURIE, S.A., FOSTER, S.J., COOPER, E.W.T. & VINCENT, A.C.J. (2004) *A guide to the identification of seahorses*. CITES/TRAFFIC, USA. (Available online: www.traffic.org/species-reports/traffic_species_fish29.pdf)

LOURIE, S.A. POLLOM, R. & FOSTER, S.J. (in review) *A global annotated checklist of the seahorses* Hippocampus *Rafinesque 1810 (Actinopterygii: Syngnathiformes): taxonomy, synonymy, biogeography and conservation status*. ZooTaxa.

ROSA, I.L., DEFAVARI, G.R., ALVES, R.R.N., et al. (2013). *Seahorses in global traditional medicine: a global overview. Chapter 10 in Animals in Traditional Folk Medicine*. Alves, R. and Rosa, I.R. (eds). Springer.

SCALES, H. (2009) *Poseidon's steed: the story of seahorses from myth to reality*. Penguin.

TESKE, P.R., HAMILTON, H., PALSBØLL, P.J., et al. (2005) *Molecular evidence for long-distance colonization in an Indo-Pacific seahorse lineage*. Marine Ecology Progress Series, 286: 249–260.

VINCENT, A.C.J. (1994) *The improbable seahorse*. National Geographic, 186: 126–140.

VINCENT, A.C.J. (1996) *The international trade in seahorses*. TRAFFIC International.

VINCENT, A.C.J, FOSTER, S.J. & KOLDEWEY, H.J., (2011) *Conservation and management of seahorses and other Syngnathidae*. Journal of Fish Biology, 78: 1681–1724.

WHITLEY, G.P. and J. ALLEN (1958) *The Seahorse and its Relatives*. Georgian House, University of California

USEFUL WEBSITES (SEE SEAHORSE AMBASSADORS SECTION FOR OTHERS)

Project Seahorse
www.projectseahorse.org
(seahorse conservation and research)

Hippocampus Info
www.hippocampusinfo.org
(reference materials)

iSeahorse
www.iseahorse.org
(citizen science records of seahorses worldwide)

Kingdom of the Seahorse
www.pbs.org/wgbh/nova/seahorse/
(BBC/Nova television documentary)

IUCN Redlist
www.iucnredlist.org
(conservation statuses and assessments)

Fused Jaw
www.fusedjaw.com
(articles about husbandry and conservation)

INDEX

INDEX

Species by scientific name [**Bold** indicates main entry]

Species by common name

Groupings

ACKNOWLEDGMENTS

This book is a compilation of many people's work, and I thank everyone who has dedicated themselves to the study of these unusual fishes. In particular, I would like to thank Amanda Vincent for her vision and perseverance in bringing seahorses to the world stage, for being so single-minded in her pursuit of their conservation, and for inspiring and leading a whole new generation of marine conservationists (myself included) to care about and protect the ocean. Thanks to the entire Project Seahorse team, past and present, and particularly to Sarah Foster and Riley Pollom, as well as Amanda Vincent, Elanor Bell, and Marie-Annick Moreau for their expert advice on earlier drafts of the text. Comments from other readers including James Maclaine (British Museum of Natural History), Neil Garrick-Maidment (The Seahorse Trust), Healy Hamilton (NatureServe) and Dave Harasti also helped improve it, for which I am grateful.

In particular, I would like to thank the editors at Ivy Press for inviting me to write this book, and for putting their faith in me at a very low point in my life. Thank you for your forbearance through all the life changes that have taken place in the last year. I have really appreciated the opportunity to work on this project. I would like to thank Denise Tackett for her inspiration and friendship, and our dives together to find seahorses, and finally my children, Erica and Oliver, for putting up with their mother's focus on these fascinating fishes, and with all the books and papers on seahorses that fill our lives and our living room.

PICTURE ACKNOWLEDGEMENTS

Cover and spine image: Getty Images/LuismiX
Back cover images: Corbis/ Bernard Radvaner, Barbara Walton/epa
Nature Picture Library/Alex Mustard, Steven David Miller
Alamy/WaterFrame
Jeffrey Jeffords/Dive Gallery.com: 2, 131.
Terri Eagle: 3.
Nature Picture Library/Alex Mustard: 4, 95, 139, 151.
Getty Images/Life On White: 6, 14, 41C, 68, 91.
Wikipedia/Asram: 11.
Richard Peters: 13, 18, 39.
Getty Images/Mike Hill: 15.
Getty Images/LuisMix: 16.
Shutterstock: 20, 23, 50, 77, 115, 145.
Corbis/Stuart Westmorland/Science Faction: 24.
Alamy/Danita Delimont: 26.
Getty Images/George Grall: 28.
Alamy/SCPhotos: 29.
Getty Images/Reinhard Dirscherl: 30.
Getty Images/Mark Newman: 31.
Getty Images/Borut Furlan: 33.
Corbis/Norbert Wu/Minden Pictures: 35.
Alamy/Papilio: 36.
Tami Weiss/Fusedjaw.com: 41BL, 105.
Ole Brett/Tropicalfavourites.com: 41T, 123.
Ülar Tikk: 41B, 113.
Sara Lourie: 43.
Tyler Stiem/Project Seahorse: 48.
Getty Images/Jerry Redfern: 48R.
Getty Images/ Andrew Watson: 51.
Sarah Foster/Project Seahorse: 53.
Nguyen Manh Ha: 55.
Alamy/Keith Poynton: 58.
Nishan Perera/Project Seahorse: 60.
Alamy/Louise Murray: 61.
Science Photo Library/Georgette Douwma: 63.
Project Seahorse: 63, 64.
Corbis/Visuals Unlimited: 75.
Science Photo Library/Scubazoo: 79.
Corbis/ Bernard Radvaner: 81.
Filip Staes: 83.
FLPA/Fred Bavendam/Minden Pictures: 85, 93.
Alamy/Auscape International Pty Ltd: 87.
Western Australian Museum: 89, 103, 138.
Nature Picture Library/Alex Mustard: 95, 151.
Silke Baron/CC BY 2.0: 97.
Corbis/Barbara Walton/epa: 99.
Alamy/cbimages: 101.
Martin Guard/eco2 Ltd., Tanzania: 102.
Bonnie Waycott: 106.
Richard Smith/Ocean Realm Images: 107.
Alamy/Noriyuki Otani: 108.
A. Diringer, "Images d'eau": 109.
Alamy/ImageBROKER: 110.
Brian Gratwicke/CC-BY-2.0: 117.
Joshua Lambus: 118.
Alamy/Arco Images GmbH: 119.
Nature Picture Library/Brandon Cole: 121.
Getty Images/Eco/UIG: 125.
Claudia Gravenstein: 127.
Alamy/WaterFrame: 128.
Getty Images/Karl Van Ginderdeuren/Buiten-beeld: 129.
Diego C. Luzzatto (CONICET): 130.
Alamy/Dray van Beeck: 132.
John E. Randall: 135.
Mårten Hansson: 140.
Alamy/FLPA: 141, 142, 155.
Image Quest Marine/Roger Steene: 143.
Claudine Lamothe: 144.
Nathan Litjens: 146.
Alamy/Nature Photographers Ltd: 147.
Alamy/cbimages: 148.
James Peake: 149.
Nature Picture Library/Steven David Miller: 152.
Paddy Ryan: 153.
Image Quest Marine/Jeff Yonover: 154.

The Western Australian Museum's Fish Collection commenced in 1896 and comprises nearly 190,000 specimens from almost 5,000 species including more than 400 primary types and thousands of secondary types. The collection is dominated by Western Australian specimens, but also includes significant marine and freshwater collections from Australia and the vast Indo-West Pacific region. The Fish Section has an active research programme in areas such as taxonomy, evolution, biodiversity and biogeography of the region's fishes.